KEEPING PACE

WITH SCIENCE AND ENGINEERING

CASE STUDIES IN ENVIRONMENTAL REGULATION

Myron F. Uman
Editor

National Academy of Engineering

NATIONAL ACADEMY PRESS
Washington, D.C.   1993

NATIONAL ACADEMY PRESS • **2101 Constitution Ave., NW** • **Washington, DC 20418**

The National Academy of Engineering was established in 1964, under the charter of the National Academy of Sciences, as a parallel organization of outstanding engineers. It is autonomous in its administration and in the selection of its members, sharing with the National Academy of Sciences the responsibility for advising the federal government. The National Academy of Engineering also sponsors engineering programs aimed at meeting national needs, encourages education and research, and recognizes the superior achievement of engineers. Dr. Robert M. White is president of the National Academy of Engineering.

Funds for the National Academy of Engineering's work on technology and environment were provided by the Andrew W. Mellon Foundation and the Academy's Technology Agenda Program. This volume consists of papers and speakers' remarks presented during a symposium entitled "Environmental Regulation: Accommodating Changing Scientific, Engineering, and Economic Understanding," held 11-12 February 1993. The interpretations and conclusions expressed in the symposium papers are those of the authors and are not presented as the views of the council, officers, or staff of the National Academy of Engineering.

Library of Congress Cataloging-in-Publication Data

Keeping pace with science and engineering : case studies in
  environmental regulation / Myron F. Uman, editor ; National Academy
  of Engineering.
     p. cm.
  Includes bibliographical references and index.
  ISBN 0-309-04938-5
  1. Environmental engineering—Case studies. I. Uman, Myron F.,
1939-   . II. National Academy of Engineering.
  TD153.K44 1993
  363.73'7—dc20   93-5530
                                                        CIP

Cover art: *Large Fringe,* courtesy of the artist, Louise H. Spindel, Falls Church, Virginia, 1993.

Printed in the United States of America

SYMPOSIUM STEERING COMMITTEE

CHARLES R. O'MELIA, *Chairman,* Professor of Environmental
Engineering, Department of Geography and Environmental
Engineering, The Johns Hopkins University
J. CLARENCE (TERRY) DAVIES, Director, Center for Risk
Management, Resources for the Future, Inc., Washington, D.C.
ROBERT C. FORNEY, Retired Executive Vice President, E. I. du Pont de
Nemours & Company, Inc., Unionville, Pennsylvania
ROGER O. McCLELLAN, D.V.M., President, Chemical Industry Institute
of Toxicology, Research Triangle Park, North Carolina
M. GRANGER MORGAN, Professor and Head, Department of
Engineering and Public Policy, Carnegie Mellon University
PAUL R. PORTNEY, Vice President and Senior Fellow, Resources for the
Future, Inc., Washington, D.C.
JOHN H. SEINFELD, Louis E. Nohl Professor and Chairman, Division of
Engineering and Applied Science, California Institute of Technology

Staff

MYRON F. UMAN, Project Officer
DEANNA J. RICHARDS, NAE Senior Program Officer
TERRIE NOBLE, Administrative Assistant

iii

Preface

Decision making for environmental regulation involves a number of steps in which scientific and engineering data are gathered and analyzed for the purpose of assessing the risks, benefits, and costs of alternative courses of action. For parties with interests in the ultimate decisions, including the regulatory authorities, other governmental bodies, and private organizations, describing the state of understanding of risks, costs, and benefits is vitally important. While technical understanding generally will not exclusively determine the outcome of the regulatory process, this information is central to assessing the risks and devising alternative mitigating strategies, if any, from which the decision-making process may choose. No one would argue that environmental regulation, if not determined by the best scientific and engineering understanding, should not at least be based on it.

In a typical case, the scientific and engineering information that is available about a particular environmental issue is vigorously debated among the parties at interest and within the government. Eventually decisions are taken, often after intense bargaining or negotiation about the applicability or interpretation of data. The process can be difficult, complex, time-consuming, and agonizing, particularly for the ultimate decision maker.

The results of the decision-making processes are regulations that serve to implement the respective laws under which authority the regulations are issued. Ideally, the regulations reflect in some way our best technical understanding at the times at which the particular decisions were made.

A fundamental characteristic of scientific and engineering knowledge, however, is that our understanding may change as new data become available. Research and development are undertaken to create new data. For example, the latest diagnostic techniques are developed to obtain ever more sensitive measurements; the results are more detailed, presumably reflecting better understanding of the occurrence of substances in the environment as well as their effects. Thus, as long as research and development continue, circumstances will inevitably arise in which the technical basis for specific regulatory decisions will be superseded by better understanding that calls into question the continued validity of the basis of earlier decisions.

Examples, taken from the daily news, abound: asbestos, chlorofluorocarbons, dioxin, lead, radon, tropospheric ozone, disinfection of drinking water, and more. In some cases, the latest information suggests that risks are less severe than previously thought; it now appears, for example, that the risks associated with exposure to asbestos dust depend on the type of asbestos, information not available at the time current regulations governing asbestos were adopted. In other cases, new data indicate that risks are greater than previously thought; such has been the case with the history of our understanding of the consequences of exposure to lead.

Many environmental laws recognize the changing nature of the technical understanding of environmental problems. Some provisions authorize research and development programs as means for improving the basis for regulation over time while others were intended by their framers to provide incentives for the development of improved, more cost-effective technology, which can also change the base of data on which regulatory decisions are founded.

The Clean Air Act, for example, provides a regular schedule for reconsidering the National Ambient Air Quality Standards, inherently assuming continuous improvement in our understanding of the sources, fates, and effects of the criteria pollutants. Several pieces of legislation incorporate technology-based provisions, such as by requiring the regulatory authorities to identify the best available or most cost-effective emissions or effluent-control technologies or the best monitoring techniques, which holders of permits must then employ.

It is apparent, however, that reconsidering earlier decisions in light of new technical understanding has proven in practice to be as daunting a challenge as the original decision making itself, if not more so. The ambient air quality standards have not been reviewed as required by statute, and evidence is scant that the various technology-based performance standards have provided incentives for innovation as their respective authors anticipated.

Why? How does the environmental regulatory system, in practice, take account of changes in technical understanding? Does experience suggest the existence of thresholds that act as barriers to reconsideration of earlier regulatory decisions? What factors influence administrative decisions to undertake reconsiderations? When might constancy be more important than revision? Does the system work as well as might be expected, or are improvements warranted?

This volume is based on a symposium organized to address these and similar questions about the ability of the regulatory system to respond to changing technical understanding of risks, benefits, and costs. The symposium was organized under the direction of a steering committee appointed by the National Academy of Engineering. Names and biographical sketches of the committee members are included elsewhere in this volume.

The symposium focused on practical experience as detailed in a series of case studies commissioned for this purpose. The introduction is based on the text of the keynote address, delivered by Robert M. White, president of the National Academy of Engineering. The bulk of this volume comprises the case studies, revised by their respective authors in light of the discussions among participants in the symposium. The volume also contains two additional essays. One, by Richard D. Morgenstern, director of the Office of Policy Analysis of the U.S. Environmental Protection Agency, is on the interplay between science and regulation. The other, summarizing lessons learned from the case studies and discussions at the symposium, was prepared by J. Clarence (Terry) Davies, a member of the steering committee. Both essays are written versions of talks presented orally at the symposium.

The case studies were selected by the steering committee to range across a broad spectrum of environmental issues and to illustrate regulatory approaches under laws dealing with air quality, water quality, safe drinking water, and hazardous substances. The objective was to draw lessons from across the particular examples to illuminate the more general issues. The case studies included examples of decision making for exposures (by-products of the disinfection of drinking water, dioxin, formaldehyde), for compliance with ambient standards (tropospheric ozone, surface water quality), and for the performance of technology (municipal waste combustion). In addition, one case study, on acid precipitation, was commissioned to examine the relationships between a dedicated federal research program and the development of regulatory policy through legislation.

The symposium, held in Washington, D.C., on 11–12 February 1993, was sponsored by the National Academy of Engineering as part of a series of activities in its program on Technology and the Environment. The steering committee identified the topics for the case studies, supervised their preparation, obtained peer reviews of the resulting papers, and organized the symposium. The committee also asked a number of people to serve as discussants for the respective papers to stimulate comments, observations, and critiques by participants at the symposium.

On behalf of our colleagues on the steering committee, we wish to extend special thanks to the case study authors and the discussants. The authors are identified with their papers. The discussants were William F. O'Keefe of the American Petroleum Institute, Louis E. Sage of the Academy of Natural Sciences of Philadelphia, B. Kent Burton of the Integrated Waste Services Association, R. Rhodes Trussel of James M. Montgomery Engineering, Frank Mirer of the United Auto Workers, and Dwain Winters of the U.S. Environmental Protection Agency.

We also wish to express our sincere appreciation to the National Academy of Engineering for its leadership and financial support and to its able program staff for facilitating our work and shouldering the administrative and editorial burdens involved in bringing the symposium and this book to life. We are specially indebted to Deanna Richards, senior program officer for technology and environment, and Terrie Noble, administrative officer for the symposium.

Charles O'Melia
Chairman, Steering Committee

Myron F. Uman
Project Officer and Editor

Contents

KEEPING PACE

WITH SCIENCE AND ENGINEERING

Keeping Pace with Science and Engineering. 1993.
Pp. 1–7. Washington, DC: National Academy Press.

Introduction: Environmental Regulation and Changing Science and Technology

Robert M. White

Accommodating environmental regulation is in many cases an engineering task that can add significant costs in the production of goods and services as it protects environmental values. How such costs are distributed is itself a contentious issue, as amply demonstrated by the estimate of costs of the recent amendments to the Clean Air Act. When environmental regulatory costs turn out in retrospect to have been unwarranted because regulatory decisions were based on inadequate or inaccurate scientific information, it's only natural to express concern, since costs will have been borne without deriving the projected environmental benefits.

The question under discussion in this volume—how does changing scientific, engineering, and economic understanding precipitate reconsideration of earlier environmental decisions?—lies behind current headlines about environmental issues. For example, the *New York Times* for Sunday, 7 February 1993, carried a thoughtful article on the proposals of the new EPA administrator, Carol Browner, to reconsider the Delaney Clause, the part of the Federal Food, Drug, and Cosmetic Act that strictly bans from food any substance that has been shown to cause cancer in laboratory animals.[1] Since its passage, the technology for measuring trace substances in food has improved tremendously, and we have become much more sophisticated in our

---

This paper was prepared in close collaboration with Myron F. Uman, who served as project officer for this symposium. Dr. Uman is assistant executive officer for special projects of the National Research Council.

understanding of risks. We now appreciate more fully that difficult trade-offs must be made among alternative risks and that some risks are not unreasonable to accept. According to newspaper reports, Ms. Browner believes that the time has come to change the Delaney Clause in light of current understanding.

In a recent editorial in *Science* magazine, Phil Abelson wrote about the phenomenon of "regulation gone amok."[2] He quoted some statistics that I found mind-boggling. He pointed out that there are 59 regulatory agencies with about 125,000 employees at work on 4,186 pending regulations. He also reported that the fastest growing component of regulatory costs is environmental regulation, which in 1991 amounted to $115 billion and is slated to grow by more than 50 percent by the year 2000.

Abelson's concern was directed not at the costs of regulation *per se*, but at the costs of regulations that, on the surface, appear to be unwarranted in light of the benefits derived. Abelson cites as an example the application of national standards for contaminants in drinking water with results that appear nonsensical under certain local conditions. One locality is being required to make large investments in equipment to remove contaminants from water that already meets the standards for those substances.

Abelson's editorial touches on the difficulty of applying national standards across a diverse country. Issues of this type arise in two of the case studies presented in this volume, one on compliance strategies for meeting ambient ozone standards in urban areas across the country, and the other on meeting ambient water quality standards in Chesapeake Bay. The conflicts between national standards and regionally variable implementation raise important issues in their own right. For current purposes, however, I only raise them to illustrate the potentially enormous implications of environmental regulations—and hence the importance and difficulty of making sound decisions.

I write of these issues as one who has been in the trenches administering regulatory processes. In one of my previous incarnations, as the first Administrator of the National Oceanic and Atmospheric Administration, I was the responsible federal official for regulation of the U.S. ocean fisheries and of other living marine resources. Although fisheries management and protection of marine mammals constituted only 10 percent of our budget, regulatory oversight and its attendant political ramifications occupied 80 percent of my time.

Regulating the take of fish or protecting porpoises and whales, like all presumably rational environmental regulation, is a matter of weighing scientific information about causes and effects against legislated criteria for regulatory action. Legislated criteria are generally an amalgam of scientific knowledge and the value judgments of our representatives in the legislature. In the real world, uncertainty clouds not only the scientific interpretation of the evidence and the economic and social costs of action but also the interpretation of the intent of the law.

Lurking, ready to pounce on decisions whichever way they go, are the affected constituencies. Depending on the intensity of public concern and the political clout of the various parties at interest, the consequences for the regulator can be unenviable, not to mention the consequences for the environment and the affected constituencies.

While the kinds of regulatory decisions that I had to make were different in some cases from the regulation of toxic emissions or effluents, we also were confronted by the infamous "mercury in fish" scare. The cause was believed to be heavy-metal discharges by industry into the oceans. In the West Coast halibut fishery, the result was the prohibition on the take of fish above a certain size because the concentration of heavy metals in fish increases with age and size. The same was true for swordfish. By examining museum specimens, however, we later found, except in certain circumstances, such as at Minamata, that the mercury in ocean fish was not a result of industrial pollution of the oceans but reflected natural levels.

The environmental case studies presented in this volume reflect the dilemma of regulation in the face of uncertain and changing scientific knowledge. I would like to probe this dilemma more thoroughly.

On the one hand, all regulatory policies promulgated legislatively or by executive actions are based on scientific and technical information that is only a snapshot of our knowledge at the time of decision. Our scientific understanding and our technologies for measurement or remediation are continuously changing, however, sometimes at a very rapid rate, unfettered and uninfluenced by the politics of the moment or constituency concerns except as politics might affect federal funding for research and development.

Regulation, on the other hand, changes slowly. This is true not only because of the ponderous nature of the legislative and regulatory process, but also because constituencies who have achieved real or imagined gains fear losing them if regulations are conformed to the latest scientific information. The institutions that must comply with regulations, such as industry, value regulatory stability highly because the overall costs of compliance can be increased by rapid changes in understanding and technology.

We are all familiar with the cases in which new understanding or technological approaches have revealed that regulation should be changed—in some cases tightened and in others loosened. Lead is the classic example in which our improved scientific understanding of health effects and our increased ability to measure ever-lower concentrations have resulted, and appropriately so, in more comprehensive regulation of lead in all human activities—in fuels, paints, pipes, dinnerware, etc. In other cases, improved understanding has revealed that perhaps earlier regulatory actions have over-estimated risks. The cases of asbestos and perhaps dioxin come to mind.

What is to be done? Regulatory actions have vast implications for human health, the quality of the environment, and economic and social welfare. We need to ask how effective are the mechanisms and procedures that have been incorporated into the regulatory process to enable it to keep pace with our scientific understanding and technological capabilities.

A host of questions face any regulatory process. Which data shall be accepted for evaluation? What endpoints are appropriate measures of harm? Who is best qualified to provide independent evaluation of the available data? What evaluation processes are the most desirable? How should the results of these evaluations be expressed?

And the process is plagued by uncertainties in our understanding of hazards, risks, costs, and benefits. In environmental regulatory affairs, we frequently are confronted with data for which neither the level of precision nor the level of accuracy is particularly high. Physicists may know the value of the speed of light to eight or nine significant digits, but in environmental affairs, we must often deal with uncertainties in the first or at best the second significant digit.

One reason that uncertainties are high is that environmental regulations address issues at the cutting edge of current scientific understanding. They address issues in which the representativeness of data and measurements are under question. As time goes on, new data are collected and help to improve our understanding and occasionally change it radically. New measurement techniques allow us to detect the presence of contaminants at lower concentrations than earlier methods did or to gauge their more subtle effects. Both the techniques and the data they produce are likely to be subject to interpretation, and there may be legitimate differences among interpretations, each with different implications for regulatory decisions.

Each party to the regulatory process therefore wants to be sure that the available data are properly reviewed and evaluated. While there may be differences among the parties in their attitudes about what constitutes proper review and evaluation, no one argues that the data ought not to be subject to this scrutiny, which presumably results in a body of technical evidence that represents the best that is available at a given time.

As I have already indicated, the body of evidence on which a regulatory decision is based—our understanding of the scientific data, engineering capabilities, and economic consequences of alternative actions—is dynamic. As a rule, scientists, engineers, and economists strive continuously to refine this understanding. As we improve in our ability to measure causes and effects, or to design alternative production or pollution abatement technologies, or to assess consequences, it is likely that sooner or later the time will come when the technical basis for a particular decision no longer represents the best available scientific, engineering, or cost data.

Some national environmental legislation recognizes the dynamic nature of the technical basis for regulation. In some cases, it mandates research and development programs to improve the data base. In other cases, it provides incentives for the development of new, more cost-effective technology. In still other cases, legislation explicitly includes schedules for reconsidering specific regulatory decisions.

The National Academies of Sciences and Engineering have been involved in such periodic reviews. In the early 1970s, when the stratospheric ozone layer was thought to be threatened by the NO_x emissions of supersonic aircraft, a substantial stratospheric research program was initiated by the Department of Transportation and several other agencies. The Academies were asked to help with a series of biennial assessments of the latest scientific knowledge of stratospheric ozone and the causes of its variability. The sequence of those reports revealed the changing nature of our understanding of the photochemistry of the stratosphere and the effects of nitrogen and chlorine on stratospheric ozone concentrations. Because our understanding of the chemical and physical processes was changing rapidly at that time, the best estimates from several successive assessments lay outside the error bars of the previous respective assessment. No wonder policymakers were confused and reluctant to act, until the so-called ozone hole was observed over Antarctica.

Another set of issues arises when the government organizes large scientific and technological research programs to improve the basis for decision making. The acid rain case is an excellent example of an attempt by the Congress through legislation to develop the necessary scientific knowledge on which to base recommendations. The National Acid Precipitation Assessment Program (NAPAP) began in the early 1980s with all the best intentions. Hundreds of millions of dollars were invested in improving understanding of the causes and consequences of acid rain. Our understanding of acid rain was greatly improved, but it is not at all clear that NAPAP provided, on a timely basis, the relevant information that Congress wanted in order to set acid precipitation policy.

It is well and good that we want to base environmental regulations on the best available technical understanding, but we need to recognize that that understanding is inherently dynamic. We need to build into the structure of the regulatory system means for reconsidering earlier decisions if and when our understanding changes sufficiently to call earlier decisions into question. On the other hand, it is impractical to attempt to revise regulations continuously in response to new, presumably better data. For one thing, the decision-making process itself is very demanding of the time, energy, and other resources of both the regulatory agency and the affected parties. Much effort, including independent reviews and often intense negotiations among the parties, goes into the effort. We cannot afford the continuous and considerable commitment of resources required.

Perhaps more important, as desirable as it is to have regulations that are based on the best, most current technical understanding, it is also desirable to have a stable regulatory regime within which the affected parties can intelligently plan to come into compliance and implement their plans. If they know what is expected of them, most firms will strive to comply. However, doing so not only takes time but also involves capital expense. Within their frameworks for planning and executing capital investments in pollution abatement technology or alternative production processes or product formulations, regulated industries prefer—and deserve—predictable regulatory regimes.

We have then two characteristics that we all would agree the environmental regulatory system should exhibit: it should keep pace with changes in our understanding of the technical aspects of the issues, and it should remain stable on a time scale sufficient for regulated parties to comply with some measure of economic efficiency. It is evident that these two normative characteristics can be, and frequently are, in conflict.

It is not unusual for scientific discoveries or economic conditions to change our understanding of the relative risks, benefits, and costs of regulation faster than industry can innovate, develop, and install improvements in pollution control or production technologies. We may not be able to predict when our technical understanding will change, but we can reasonably predict that occasions will indeed arise in which it changes much more quickly than we might have thought—and more quickly than can be readily accommodated by affected parties.

How quickly or on what time scale should we either expect or accept changes in regulation as a consequence of new scientific, engineering, or economic understanding? Is there a threshold for change, an accumulation of new understanding which, when reached, should trigger a response by the regulatory system? When is the legitimate desire of the regulated industry for stability an appropriate barrier to change?

The primary question of policy is this: Does the current environmental regulatory system strike an appropriate balance between dynamic change and stability? As my colleagues and I thought about this question, we discovered to our surprise that while the regulatory combatants have "war stories" to tell, there have been few, if any, attempts to analyze how the system has been working in practice.

We know, for example, that the national ambient air quality standards for the criteria pollutants have not been reviewed on the five-year cycle that is mandated in the Clean Air Act. Why? Is it because new data or improved technical understanding has not developed on this time scale, or because the affected industries need longer periods over which the standards remain constant, or because the investments of time, personnel, and other resources—including political capital—are too great to repeat the decision-making process this frequently? Should this provision of the Clean Air Act be changed?

To take another example, several pieces of environmental legislation contain provisions that, at the time they were adopted, were intended to provide incentives for innovation in pollution control technology. The development and demonstration of improved technology were thought to compel reconsideration of performance standards for abatement, at least in some cases. Did the provisions work as intended?

When many of these provisions were enacted, the technological focus in environmental affairs was on pollution control, that is, on end-of-pipe treatment. More recently, the focus has been on pollution prevention, that is, on product reformulation or process redesign. How does this change in focus affect regulatory decision making on performance standards? Are changes warranted in the provisions themselves?

This volume is about keeping pace with changing technical understanding in environmental regulation. Its purposes are to shine a spotlight on the competing demands for keeping regulations in step with current knowledge and for maintaining regulatory stability and to serve as a catalyst for further consideration of and debate about the appropriate balance between these goals. My hope is that the case studies and essays in this volume will stimulate additional analytic examination of current policies and past practices, leading to a better understanding of whether the current regulatory system does about as well as can be expected or, if not, what alternatives might be considered.

NOTES

1. Keith Schneider. A trace of pesticide, an accepted risk. The New York Times, 7 February 1993.

2. Philip H. Abelson. Regulatory risks. Science 259(8 January 1993):159.

Keeping Pace with Science and Engineering. 1993.
Pp. 8–38. Washington, DC: National Academy Press.

Nutrient Loadings to Surface Waters: Chesapeake Bay Case Study

Thomas C. Malone, Walter Boynton, Tom Horton,
and Court Stevenson

Nutrient pollution poses the greatest of all recognized threats to Chesapeake Bay.

L. Eugene Cronin, *Baltimore Sun*, March 22, 1967

The thing that really bothers me is that when people like me grow old and die off, there leaves a generation back that has no idea of what the conditions of the river were. They don't have the memory at all about the ten barrels of crabs a day a person could catch . . . about the soft crabs crawlin' in the clear water across grassy bottoms. . . . There's going to be nothing in those computer memory banks . . . that can generate the enthusiasm for the Bay that those sights and sounds did.

Senator Bernie Fowler, *Baltimore Sun*, June 14, 1992

Nutrient inputs that result from human activities often cause aquatic ecosystems to become overloaded with nutrients and deficient in oxygen, a process referred to as cultural eutrophication. This phenomenon occurs when nutrient inputs exceed the ability of the system to absorb and use them—its assimilation capacity—resulting in the degradation of water quality.[1] Since the 1960s, environmental scientists and managers have struggled with the causes, consequences, and prevention of eutrophication. Our analysis

8

is concerned with the relationship between environmental research by the science community and the formulation, implementation, and evaluation of nutrient control strategies by the management community.[2] We will not explicitly treat such important problems as water use, the enforcement of government regulations, or the development of a new social ethic for the public stewardship of natural resources. We ask the question, "How and why does the management community respond (or not respond) to new scientific information on the causes and consequences of nutrient loadings to surface waters?"

Since the flow of information between the research and management communities is neither one way nor linear, we must also be concerned with the response of the research community to the needs of management. The interplay among the research and management communities characteristically involves feedbacks between different levels of government (local, state, and federal), public and private institutions, citizens' groups, and individuals. The complex nature of these interactions and the current compartmentalization of ecology and economics into opposing forces create an inertia that reflects both the bureaucracy within which the research and management communities are imbedded and the multiple ecological, economic, and social interests that management agencies represent.

For this case study, we have selected the Chesapeake Bay. As for most of the nation's coastal ecosystems, nutrient loading to the watersheds of the main Bay and its tributaries (Figure 1) has increased substantially in the decades since World War II, largely as a consequence of rapid population growth and increases in agricultural fertilization, the density of farm animals, and atmospheric inputs. This has been a matter of increasing concern throughout the Chesapeake Bay watershed, especially in the states of Maryland and Virginia, the economies of which are closely tied to the Bay and its resources. Perhaps as a consequence of this and its proximity to Washington, D.C., the Bay has been the subject of much political and scientific attention and controversy since the early 1960s. For these reasons, and because the responsibility for nutrient management resides with individual states, our analysis of the relationship between science and management will focus on the state of Maryland. We hope to show how uncertainty, the availability of cost-effective solutions, and forces inherent to the conduct of the science and management communities have interacted to (1) limit the information exchange critical to the objectives of both communities and (2) inhibit the timely development and implementation of comprehensive nutrient management strategies.

The environmental effects of anthropogenic nutrient[3] enrichment (cultural eutrophication) began to receive national and international attention in the 1960s with major efforts to control nutrient loadings and continued during the 1970s to the present. In the Chesapeake region, the main event

FIGURE 1   Drainage basin of Chesapeake Bay.

during this period was the U.S. Environmental Protection Agency (EPA) Chesapeake Bay Study mandated by Congress in 1976, implemented in 1977, and completed in 1983 with the release of the Chesapeake Bay Program reports: *A Profile of Environmental Change* (EPA, 1983d), *Findings and Recommendations* (EPA, 1983b), and *A Framework for Action* (EPA, 1983c). The implementation of this study and the publicity that surrounded its completion had a major impact on the perspectives of both science and management communities and on the interplay between them, much of which was (and is) modulated by public interest and political pressure. Thus, for the purposes of our analysis, we divide our narrative of the sequence of events into the "formative" years prior to the EPA Bay Study (1965–1977),

the period of the EPA Bay Study (1977–1983), and the "action" years following the Bay Study (1983–1992) (see Figure 2).

THE FORMATIVE YEARS

Nationally, the perception of eutrophication as a water quality problem
was largely based on studies of the effects of nutrient loading to freshwater
systems in which phosphorus (P) is usually the controlling nutrient (American Society of Limnology and Oceanography, 1972; National Research Council,
1969). Vollenweider (1968, 1976) published his widely accepted model of
phosphorus limitation in lakes, an empirical analysis that also appeared to
be applicable in concept to estuaries where marine and freshwaters mix
(Ketchum, 1969). The generality of Vollenweider's model for lake systems
was vividly demonstrated through the experimental manipulation of lakes in
Canada (Schindler, 1974). Schindler (1977) went on to show that lake
communities are able to compensate for deficiencies of nitrogen (N) and
carbon through gaseous exchange with the atmosphere, and that attempts to
control nitrogen loading may actually degrade water quality because they
may result in the growth of noxious blue-green algae (which are capable of
fixing nitrogen). In contrast, research in marine systems was beginning to
produce evidence that N, not P, is the principal nutrient limiting primary
production (Ryther and Dunstan, 1971). However, despite new scientific
evidence that N-control would also be necessary (see Boynton et al., 1982;
Nixon and Pilson, 1983), nutrient management in the Chesapeake region
through the 1970s and into the 1980s was dominated by the growing body
of evidence for phosphorus limitation in freshwater systems.

Federal Studies and Legislation

The Water Pollution Control Acts (also known as the Clean Water Acts,
CWAs) of 1965 and 1972 reflected a growing concern over the pollution of
lakes and rivers and the threat this posed to the nation's water supply, living
resources, recreational use, and aesthetics (see Figure 2). The 1965 CWA
required the adoption of enforceable ambient water quality standards for all
interstate waters. As in the past, the primary responsibility for nutrient
management was vested in the states. In the 1972 amendments to the CWA,
Congress drastically altered the nation's management approach. It changed
the focus from ambient water quality to effluent standards by calling for the
nationwide implementation of secondary treatment. Technology-based performance standards became the basis of regulating nutrient (and other contaminant) inputs, and federal funding to the states for the construction and
upgrading of sewage treatment plants (STPs) was increased from 55 percent

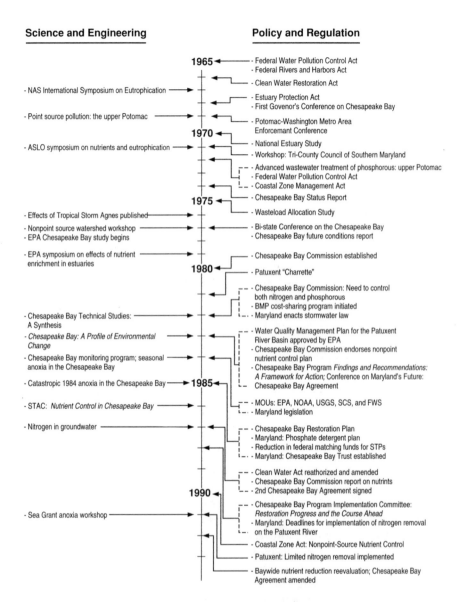

FIGURE 2   Timeline of significant events in the Chesapeake Bay management program.

to 75 percent of capital costs. The act also outlawed all point-source discharges of contaminants and established a permit process for dischargers who could not meet this requirement. This was the National Pollutant Discharge Elimination System (NPDES), which set legal limits on the quantities of contaminants that could be discharged. The 1972 CWA effectively gave the federal government the enforcement power to regulate nutrient inputs to the nation's surface waters. The responsibility for implementation remained with the states, which were mandated to report on water quality within their borders beginning in 1975. Stimulated by the availability of federal funds and guided by the prevailing "wisdom" calling for the control of point-source P inputs, a nationwide effort was set in motion to upgrade all STPs to secondary treatment, with advanced wastewater treatment for removing phosphorus as necessary.

In addition to the CWAs, several studies were initiated by federal legislation during this period. The 1965 Rivers and Harbors Act directed the U.S. Army Corps of Engineers to conduct a comprehensive "study of water utilization and control of the Chesapeake Bay Basin," including water quality control. The 1966 Clean Water Restoration Act directed the Department of the Interior to conduct a study of estuarine pollution nationwide, and the 1968 Estuary Protection Act directed Interior to "study and develop the means to protect, conserve, and restore" the nation's estuaries.

This legislation resulted in four important reports, which laid the foundations for and ultimately led to the EPA Bay Study:

1. In 1969 the Water Pollution Control Administration reported on the adverse effects of nutrient enrichment in the tidal freshwater reaches of the Potomac and Patuxent rivers.

2. In 1970 the Interior Department's national estuarine study, conducted by the Fish and Wildlife Service, recommended that "An all-out cleanup program for the Chesapeake Bay area might serve as a national and even an international demonstration area, showing what can be accomplished by an enlightened public and a responsible Congress."

3. In 1973 the Corps of Engineers released its Chesapeake Bay Status Report in which water quality in the Bay was assessed as good, with local problems limited to the tidal freshwater reaches of some of the Bay's tributaries.

4. In 1977 the Corps presented its *Chesapeake Bay: Future Conditions* report (published in 1978) to the bi-state conference on the Bay. The report acknowledged the *potential* significance of excess nitrogen and phosphorus loading, listed (but did not quantify) major nutrient sources, and suggested that land use and nonpoint sources of nutrients are related.

It is noteworthy that, although the Corps and Interior reports acknowledged the link between land use and nonpoint-source nutrient loadings, the management community would not give control of nonpoint sources serious attention until the late 1980s and early 1990s. This preoccupation with point sources is evident in EPA's 1975 report to Congress, which proclaimed overenrichment from sewage to be a major problem in the nation's estuaries. The Chesapeake Bay was identified as being particularly vulnerable. Under the leadership of Maryland's Senator Charles McC. Mathias, this would cause the Congress in 1976 to direct the EPA to "undertake a comprehensive study of the Bay's resources and water quality, and to identify appropriate management strategies to protect this national resource."

The Chesapeake Region

In the midst of these studies and federal legislation, symptoms of overenrichment were appearing in Chesapeake Bay and its tributaries during the late 1960s and early 1970s. Massive algal blooms, oxygen depletion, and fish kills in the upper Potomac River were gaining the attention of the public and federal government officials in Washington, D.C. Scientists raised the issue of excess nutrient enrichment in general and N loading in particular during the first Governor's Conference on the Chesapeake Bay in 1968 (Jaworski, 1990). Nutrient distributions and historical records dating back to the 1930s indicated a trend toward increasing eutrophication in the upper reaches of the Bay and its tributaries (Carpenter et al., 1969; Heinle et al., 1970). Declines in the abundance of submerged aquatic vegetation (SAV) were documented in the Rhode River estuary (Southwick and Pine, 1975), the upper Patuxent River estuary, and the main Bay (Bayley et al., 1968, 1978). Stevenson and Confer suggested (1978) that these declines might be related to decreased light because of excessive algal growth. Evidence was also accumulating that wastewater inputs to the upper Patuxent River were beginning to cause eutrophication in the lower Patuxent (Flemer et al., 1969). The 1975 Wasteload Allocation Study, conducted by Hydroscience, Inc. under contract to the state of Maryland, concluded that P is the primary nutrient limiting phytoplankton production in the Bay and that the removal of P from sewage wastes is the highest priority for improving water quality. At the same time, research on estuarine circulation highlighted the need for a more systemwide approach to material transport and retention (e.g. Heinle et al., 1970; Pritchard, 1969).

The concerns of federal officials, scientists, and some local officials are clearly documented by the *Baltimore Sun*. For example, U.S. Congressman Carlton Sickles from Maryland claimed that the Bay is polluted "to the point of public danger," and an official of the U.S. Fish and Wildlife Service reported that the Bay "could be dead in five years" (April 23, 1966). The Assistant Secretary of Interior for fish and wildlife concluded that "you can't clean the Bay up, you've got to clean up the watershed" (August 17,

1966). Congressman Rodgers Morton expressed concern that the Bay is getting worse (February 17, 1967), and the founders of the Chesapeake Bay Foundation charged that "ecologically, the whole Bay is in danger" (June 19, 1967). Leading Chesapeake Bay scientists announced that "Nutrient pollution poses the greatest of all recognized threats to Chesapeake Bay" (L. Eugene Cronin, March 22, 1967) and that concerns over thermal pollution from power plants are distracting the science and management communities from the real problem, sewage pollution (Donald Pritchard, July 1, 1970). During this period, a study released by the Baltimore Regional Planning Council concluded that excess N and P inputs from sewage, agriculture, and natural sources were among the Bay's most important pollution problems (November 5, 1968).

In contrast to the perspective of federal officials and reports by local scientists and citizens' groups, state officials in Maryland insisted that the Bay was doing just fine. A representative of the State Board of Natural Resources referred to claims that the Bay is polluted and a public hazard as "irresponsible" (April 23, 1966). The Maryland Department of Chesapeake Bay Affairs issued a statement that "Bay water quality is good and getting better" (June 20, 1969), and Governor Marvin Mandel announced that "water quality rivals that of 25 years ago" (June 25, 1969). As late as 1977, state management officials continued to claim that the Bay was healthy and that, with the exception of a few hot spots, changes in water quality were due to natural climatic cycles (February, 1977, *Baltimore Sun* series, "Chesapeake Still at Bay"). Thus, the governing body responsible for implementing nutrient control plans, the state, was the least receptive to scientific evidence indicating the early stages of baywide eutrophication.

Control of Point Source Nutrient Loading

In the late 1960s, Jaworski et al. (1969, 1972) documented long-term nutrient trends and related changes in the ecology of the upper, fresh reach of the Potomac. For the first time in the Chesapeake region, Jaworski et al. clearly demonstrated a relationship between nitrogen and phosphorus loading from municipal wastewater discharges and deteriorating water quality, and prescribed a program of advanced wastewater treatment to remove N and P, and lower biological oxygen demand (BOD), a measure of nutrient loading. In 1969 the Potomac River-Washington Metropolitan Area Enforcement Conference agreed to set limits on the amounts of P and N that could be discharged into the upper estuary from STPs as well as on BOD levels (Jaworski, 1990). The agreement was achieved in part because the Washington metropolitan area was faced with a ban on new construction if no action was taken and in part because President Johnson, upon signing the 1965 CWA, made restoration of the Potomac a national priority. Jaworski's research provided the scientific basis for action, but the politics of the day provided the leverage. By 1972 the Blue Plains STP, which discharges into the tidal (freshwater) Potomac, had begun construction of an advanced wastewater treatment facility to remove P and lower BOD. Implementation of N removal was delayed, in part because the management community was skeptical of the need and in part because there was no cost-effective technology.

As this chapter in the Potomac episode was drawing to an end, a grass roots confrontation was developing in the Patuxent River watershed. It involved local politicians and scientists on one hand and regulatory agencies of the state and federal governments on the other (Bunker and Hodge, 1982). In 1971 a workshop involving university scientists and the Tri-County Council of Southern Maryland concluded that the water quality of the lower (salty) Patuxent River estuary had declined to unacceptable levels as a consequence of increases in municipal wastewater nutrient loadings to the upper (fresh) Patuxent. Critical to this conclusion was the existence of "baseline" water quality data collected in the 1930s by university scientists. Armed with this information and a commitment to restoring the Patuxent, the Tri-County Council under the leadership of Senator Bernie Fowler appealed to the state for action over the next five years (1972–1976) to no avail. Finally, in 1977, the council filed suit against the EPA to halt the expansion of an upstream STP until an environmental impact statement could be prepared. In 1978 the council again filed suit, this time against both the state and EPA, claiming that the Patuxent River Basin Water Quality Management Plan, which had been approved by EPA, violated 13 of 15 requirements of the 1972 Clean Water Act. The plan advocated P control as the preferred advanced wastewater treatment method for controlling eutrophication of the Patuxent. The council felt that N control was also needed, a position advocated by the Patuxent River Technical Advisory Group (TAG), an ad hoc committee of prominent university scientists.

The U.S. District Court ruled in favor of the Tri-County Council and directed EPA to prepare an environmental impact statement and the state and EPA to prepare a new water quality plan for the basin. As part of this process, the state contracted with HydroQual, Inc. to assess the impact of a set of nutrient control scenarios using a computer model. The model predicted that P removal would be sufficient, a conclusion that the TAG did not agree with. Following an evaluation of the HydroQual model, the TAG concluded in a letter to William Eichbaum (assistant secretary for the newly created Office of Environmental Programs, Maryland Department of Health and Mental Hygiene, February 6, 1981) that, although the model "is at or near the state-of-the-art for water quality modeling," uncertainties associated with the entire modeling process "preclude the use of model projections as the sole foundation for a management decision of this nature."

At this point the state was in a bind. In the absence of an approved nutrient control plan, the federal government was threatening to withdraw funding to build and upgrade STPs on the river. Faced with the loss of millions of dollars, the state sponsored the Patuxent "Charrette" in 1981, a historic conference organized by Mr. Eichbaum. Using a time-constrained, conflict-resolution process to reach a consensus, the stalemate was broken, laying the foundations for the Patuxent River Nutrient Control Plan for controlling both point and nonpoint inputs of N and P. Like the Potomac plan, the Patuxent plan set limits on total N and P loadings to the river as a whole, and, again, economics was an important factor. Unlike the Potomac plan, which was restricted to the tidal freshwater reach of the system and was formulated quickly in response to new scientific information, the Patuxent plan was truly basinwide and took a decade of struggle and confrontation to develop.

Control of Nonpoint Source Nutrient Loading

A major event occurred in June 1972 that would have a delayed but dramatic impact on the subsequent course of nutrient research and management throughout the Chesapeake region. Hurricane Agnes arrived, inundating the watershed with up to 18 inches of rainfall.[4] The watershed as a whole (64,000 square miles) received over 5 inches in less than three days. Agnes served as a "lightning rod," focusing research activities on a number of important questions, including the response of the Bay and its tributaries to nutrient enrichment (Cheaspeake Research Consortium, 1976). (The immense amount of water runoff carried with it large amounts of nutrients from nonpoint sources such as fertilizers and animal wastes.) The storm demonstrated the systemwide susceptibility of the Bay to nutrient enrichment. Major findings included large increases in nutrient levels caused by high runoff and erosion, and the realization that most of the large quantities of nutrients delivered to the Bay are retained within the Bay (rather than being exported to the ocean). Much of the nutrient input entered the sediments and was released during subsequent years, resulting in unusually high phytoplankton production (Boynton et al., 1982). In effect, Agnes brought important environmental issues before the public and primed the science and management communities for what was to become the EPA's Chesapeake Bay Study.

Clark et al. (1973) made an early assessment of nonpoint nutrient inputs in the Chesapeake watershed. They reported that N runoff from agriculture was more than an order of magnitude higher than that from forested areas. These results were reflected in Maryland's 1975 report to EPA (as required by the 1972 CWA), which emphasized point source inputs but also acknowledged that, "The heavy use of fertilizers and manure on the land results in some runoff to the streams." In 1977 the National Science Foundation

(NSF) sponsored a major workshop on watershed research in North America at the Smithsonian Center for Chesapeake Bay Research in Edgewater, Maryland. Results presented at the workshop confirmed that nonpoint N inputs were a major, if not the dominant, term in the N budget of the Bay. Although managers from both state and federal agencies attended the watershed workshop, more than a decade would pass before this reality would begin to be incorporated into a management scheme specifically directed at nutrient control. There was strong resistance by the management community in general, and by agricultural interests (both scientists and managers) in particular, to the idea that farming practices are related to nutrient loading and water quality in the Bay. This resistance was expressed by the Secretary of the Maryland Department of Natural Resources (DNR), James B. Coulter, who in referring to nonpoint nutrient sources, was quoted as stating that, "There is more alarm than is necessary; it can be controlled with just good housekeeping and old fashioned general sanitation" (*Baltimore Sun*, February 7, 1977).

Unfortunately, this "common sense" approach relied heavily on best management practices (BMP), which were intended primarily to minimize the loss of soil and thereby increase or sustain agricultural productivity. Because most P enters the estuary attached to particles, one by-product of this strategy has been to reduce nonpoint inputs of P. Despite earlier warnings (Walter et al., 1979) and information on loading rates (Jaworski, 1981) that indicated that BMPs derived from soil conservation would have little impact on dissolved nutrients such as N, management planning continued to stress problems of erosion, with little consideration for nutrient control per se. The significance of this was highlighted by studies in the Choptank River basin (on the eastern shore of the Bay), which indicated that nonpoint sources account for about 80 percent of N and 60 percent of P inputs (Lomax and Stevenson, 1981). The emphasis on point-source nutrient control would not begin to change until after the release of the results and conclusions from the Bay Study in 1982 and 1983.

The fact that the Bay is imbedded in a large watershed (about 28 units of land area for each unit of Bay surface area), which was being rapidly modified by human activities, was not generally a part of management or scientific thinking at the time. Management was focused on point-source discharges, and funding for research tended to focus the science community on the effects of sewage and thermal discharges. The problems of overenrichment were thought to be restricted to a few local tributaries such as the upper Potomac and Patuxent River estuaries, where point-source inputs were clearly related to the degradation of water quality. Despite the effects of Tropical Storm Agnes and subsequent research findings, the baywide impacts of nonpoint nutrient loading were not broadly appreciated at this time. Agnes planted the seeds, but serious attempts to understand and control nonpoint sources would await the completion of the EPA Bay Study, the development of a comprehensive watershed (multistate) approach, and the results of research in the 1980s that would document the links between agricultural practices and nutrient loading.

THE PERIOD OF THE EPA BAY STUDY

Setting the Stage

Largely in response to baywide declines in the abundance and harvest of living resources (e.g., submerged aquatic vegetation [SAV], oysters, and shad), Congress in 1976 directed the EPA to conduct a comprehensive, systemwide study of the resources and water quality of Chesapeake Bay and to recommend management plans to protect and restore this national resource. At the same time, the Chesapeake Research Consortium (CRC, formed in 1974 to facilitate a coordinated, baywide research effort) had begun planning for a Maryland-Virginia, bi-state conference on the Bay. Stimulated in part by the *Chesapeake Bay Existing Conditions Report*, released by the Corps of Engineers in 1973, and the Corps's 1977 *Future Conditions Report* (presented at the conference), leading scientists and managers met to discuss the major environmental issues of the day. Secretary Coulter opened the 1977 conference by describing the Chesapeake Bay as "a beautiful, productive body of water that provides a satisfying livelihood to many persons and boundless pleasure to many others." He went on to caution the scientific community that "if, in our zeal to sell a program or carry a point of view we place in the public's mind a picture of a dirty Bay, a Bay that is a threat to the health of fish and man alike, we will do great and needless harm."

It was in this context that leading scientists discussed the Bay environment and concluded that anthropogenic nutrient inputs were the most serious threat to the health of the Bay and acknowledged once again the importance of nonpoint nutrient inputs. Consensus among scientists and managers could only be reached on two broad issues. First, the underlying causes of declines in living resources were uncertain; and second, there was a need for a single government entity to oversee the restoration of the Bay.

In 1977, drawing to a great extent on recommendations of the bi-state conference, the EPA initiated a five-year study emphasizing the problems of nutrient enrichment, toxic substances, and the decline of SAV. As part of this study, the EPA funded CRC in 1979 to organize an international symposium on the *effects of nutrient enrichment in estuaries* (Neilson and Cronin, 1981). Research presented at the symposium highlighted the causes and consequences of nutrient loading in estuaries. Of particular significance for the Chesapeake Bay was the presentation of the first good estimates of total nutrient loads to the Bay, which showed the quantitative importance of nonpoint sources (Jaworski, 1981). When considered in the context of the NSF-sponsored workshop on watersheds two years before, the results of this symposium sent a strong signal that nonpoint sources of nutrients would have to be considered as part of nutrient management plans.

Recognizing the need for interstate, basinwide planning, joint legislation by the Maryland and Virginia general assemblies created the Chesapeake Bay Commission in 1980 to coordinate management activities and advise the legislators of both states. In 1982 the commission acknowledged the need to control both N and P inputs. It worked to formulate, and secure support for, legislation that in 1984 would facilitate the implementation of a broad range of nutrient management actions. In 1983 the commission also recognized the need for a watershed approach when it endorsed the Patuxent River Basin plan as a model for a comprehensive nutrient control strategy that would address the control of both point and nonpoint loadings in terms of total inputs. In the years following the completion of the Bay Study, the commission continued to be an important forum for promoting and guiding legislative actions, as well as the 1983, 1987, and 1992 Bay Agreements.

The Bay Study

The EPA Bay Study involved some 50 research projects, the results of which are summarized in *Chesapeake Bay Technical Studies: A Synthesis* (EPA, 1982) and in *Chesapeake Bay: A Profile of Environmental Change* (EPA, 1983d). These reports supported earlier speculation that water quality was deteriorating, that many living resources were declining, and that these changes were related in some way to land use in the drainage basin. The reports presented evidence based on data collected between 1950 and 1980, that nutrient and chlorophyll concentrations were increasing and that these increases might be related to the baywide decline in SAV and to summer oxygen depletion in the main Bay. It was suggested that declines in fisheries might be related to deteriorating water quality, especially in the upper and midbay and in the upper reaches of the western tributaries. A baywide analysis of nutrient inputs also confirmed the importance of nonpoint sources, which were estimated at the time to supply about 65 percent of P and 80 percent of N inputs. The widespread decline in the abundance of SAV in the Bay, initially described by Stevenson and Confer (1978) and confirmed by Orth and Moore (1983), was shown to be primarily a consequence of nutrient enrichment (Kemp et al., 1983; Twilley et al., 1985). Results from the Bay Study suggested that the whole Bay was changing and that many of these changes were related to increases in nutrient inputs from municipal wastes and agricultural runoff.

Results from the SAV component of the study contributed more to the understanding of the consequences of nutrient enrichment than did the nutrient enrichment component itself. Overall, little new information was generated that would further a mechanistic understanding of the causes and consequences of point- and nonpoint-source nutrient loadings. Trends in water quality parameters were not statistically well documented and advances in the understanding of underlying causes of eutrophication were limited. With the notable exception of the nutrient-SAV work, little was learned that would allow a cause-effect, quantitative analysis of the relationships between nutrient inputs, water quality, and living resources in the main Bay. In a qualitative way, these links made sense, but the scientific evidence needed to make the case remained weak.

With the publication of *Chesapeake Bay: A Framework for Action* in 1983, the EPA presented its recommendations for a range of actions to restore the Bay. Despite the lack of scientific information needed to quantify the effects of anthropogenic nutrient inputs, the major focus of these recommendations was on the control and monitoring of nutrients "to reduce point and nonpoint source nutrient loadings to attain nutrient and dissolved oxygen concentrations necessary to support the living resources of the Bay."

Following the release of this report, the Citizen's Program for the Chesapeake Bay (precursor of the Alliance for the Chesapeake Bay) organized a conference, "Choices for the Chesapeake: An Action Agenda," which laid the foundation for the subsequent signing of the 1983 Chesapeake Bay Agreement by the governors of Maryland, Virginia, and Pennsylvania, the mayor of the District of Columbia, the administrator of the EPA, and the chairman of the Chesapeake Bay Commission. This historic agreement committed the EPA and the states "to improve and protect the water quality and living resources of the Bay system, to accommodate growth in an environmentally sound manner, to ensure a continuing process of public participation, and to facilitate regional cooperation in the management of the Bay." An administrative structure was created to achieve these goals. It consisted of the Chesapeake Executive Council (the appropriate cabinet designees of the governors, the mayor of the District of Columbia, and the regional administrator of EPA), a Citizens Advisory Committee, a Science and Technical Advisory Committee, and an Implementation Committee. For the first time, the states and EPA officially admitted that systemwide problems existed and that they were getting worse, not better, under existing management practices. The momentum created by the 1983 Bay Agreement led to a flurry of legislative actions in 1984, the establishment of the Chesapeake Bay monitoring program, and an unprecedented decade of nutrient-related studies by the science community.

Public participation was an essential ingredient that helped to sustain the high level of environmental activity and facilitated a somewhat reluctant interaction among scientists and managers following the 1983 Bay Agreement. The Alliance for the Chesapeake Bay, which staffs the Citizens Advisory Committee to the Chesapeake Program, was formed to work toward the restoration of the Bay through public education, dissemination of information (e.g., the *Bay Journal*), and citizen involvement. Environmental advocacy groups grew rapidly in membership and influence. The Chesapeake Bay Foundation (CBF), with its motto of "Save the Bay" and a fund-raising pitch that typically begins with "the Bay is dying," increased its membership from about 10,000 in 1983 to more than 80,000 in 1992, making it the largest regional, nonprofit environmental organization in the nation. This allowed CBF to mobilize public opinion and apply pressure on the government to continue the course established by the Bay Study and subsequent agreements.

THE ACTION YEARS

For many, 1984 was considered to be the year of the Bay. At the federal level, EPA, the National Oceanic and Atmospheric Administration (NOAA), the Geological Survey, Soil Conservation Service, and the Fish and Wildlife Service signed memoranda of understanding to coordinate research and management activities related to environmental issues. No single issue preoccupied the Chesapeake Bay's environmental agenda during the 1980s and early 1990s more than the effects of excessive nutrient inputs from point and nonpoint sources. This was reflected in the level of legislative activity in 1984 and 1985, the 1987 and 1992 Chesapeake Bay Agreements, and continuing debates between and within the management and science communities concerning issues such as the need to control both N and P inputs, methods for controlling nonpoint sources of N, and the need for baywide versus basin-specific nutrient control strategies.

Among the most important legislative acts of the 1984 session of the Maryland Assembly was the appropriation of funds for a comprehensive water quality monitoring program. Responding to recommendations in *Chesapeake Bay: A Framework for Action*, EPA (1983c) and the state of Maryland established the most comprehensive water quality monitoring program ever to be implemented in an estuarine system. The Chesapeake Bay Monitoring Program addressed a major problem encountered during the years of the Bay Study—the inability to document how the Bay had changed. Temporal and spatial variability would be monitored in order to determine long-term trends in water quality and living resources, to resolve natural cycles and anthropogenic sources of variability, and to evaluate the efficacy of pollution control programs.

Nutrient Research

The Bay Study spawned an unprecedented research effort during the 1980s, which focused on five key issues: (1) the significance of the benthos (i.e., the bottom of the Bay) in the nutrient dynamics of the Bay, (2) the relative importance of nitrogen and phosphorus in limiting phytoplankton production, (3) quantification of SAV responses to changes in N and P concentrations in the Bay, (4) the causes and consequences of nutrient loading in terms of oxygen depletion (habitat loss) and its impact on living resources, and (5) the significance of and methods for controlling nonpoint-source nitrogen inputs. Research on nutrient fluxes from the benthos demonstrated the role of benthic-water column interactions in controlling the nutrient dynamics in shallow estuaries (Boynton et al., 1980; Kemp and Boynton, 1984). The work of Officer et al. (1984) and Seliger et al. (1985) highlighted seasonal depletion of dissolved oxygen as a measure of the Bay's capacity to support living resources and of climatic variability in controlling the Bay's response to nutrient inputs. Nutrient enrichment studies (e.g., D'Elia, 1987; D'Elia et al., 1986) provided the basis for a report released by the Scientific and Technical Advisory Committee of the Chesapeake Bay Program in 1986 presenting clear and compelling scientific evidence that both P and N removal would be required to improve water quality in the Bay and its tributaries. The report emphasized the fact that cost-efficient technologies are now available for the combined removal of P and N, and reducing BOD, and strongly recommended that N removal be implemented.

Multidisciplinary research on the Bay during the 1980s led to a 1991 workshop sponsored by the Maryland and Virginia Sea Grant College Programs and to the release of a report in 1992 entitled *Dissolved Oxygen in the Chesapeake Bay: A Scientific Consensus.* Based on a comprehensive analysis of oxygen dynamics in the main Bay (Smith et al., 1992), the report emphasized the susceptibility of the Bay to seasonal oxygen depletion and to climatic variability, and concluded that nonpoint nutrient inputs were the primary sources of the nutrients that fueled oxygen depletion in the main Bay. The report also endorsed the goal of achieving at least a 40 percent reduction in nutrient inputs to the Bay and underscored how little is known concerning the relationship between water quality and the capacity of the Bay to support living resources.

In the same year, the results of statistical analyses of monitoring data from the main Bay for 1984–1990 showed a significant decline in total phosphorus (19 percent), a small but significant rise in total nitrogen (2 percent), and no significant trend in oxygen depletion in bottom water (EPA, 1992). The decline in P apparently reflects the effectiveness of point-source controls (enhanced P removal by STPs and the phosphate ban enacted by the Maryland General Assembly in 1985). The relatively small change in total nitrogen levels suggests that the achievement of a 40 percent reduction in N input will depend on the success of nonpoint source controls of N, a conclusion that is consistent with results from the Choptank River where nonpoint sources dominate nutrient inputs and the concentrations of N and phytoplankton biomass increased from 1985 to 1991 (Stevenson et al., 1993). In addition, research on the movement of N through the watershed demonstrated that the major route of N loss from agriculture systems occurs through groundwater (Staver et al., 1987), indicating that effective control of N inputs to the Bay must address subsurface water movements (Staver et al., 1989).

Nutrient Management

Activity in the management arena was also stimulated by the Bay Study. In 1984 the Maryland General Assembly enacted eight authorization and assistance bills that contributed to the management of nutrient inputs to the Bay. Bills aimed at point source management included (1) the State Financial Assistance Program, which created a water pollution control fund and established policy and procedures for using these funds to assist local governments in constructing STPs and implementing stormwater management programs (and to encourage farmers to implement BMPs); (2) the Water Quality Loan and (3) Existing Loan Authorizations bills, which provided bond authorization and increases in the state's share of STP construction costs (in anticipation of reductions in federal funding from 75 to 55 percent) so that the cost to local governments would remain at 12.5 percent; and (4) the Water Pollution Control bill, which provided the authority to require and enforce pretreatment of industrial wastes. Additional funding for the Bay restoration effort was made possible in 1985 when the Maryland Assembly created the Chesapeake Bay Trust to support private and corporate involvement through private donations.

Despite the cumulative evidence that N removal was needed to improve water quality (from the Patuxent Charrette in 1981 to the 1986 STAC report and the 1987 Chesapeake Bay Agreement), resistance within the management community to implementing the measures needed to reduce N inputs remained strong through most of the 1980s. With respect to point sources, management officials in Maryland as late as 1983 held the point of view that nitrogen removal in "all treatment plants in the State that discharge" to the Chesapeake Bay should "never" be required (memoranda from technical and permit staff to the director of the Maryland Water Management Administration dated May 30 and July 20, 1983). Ultimately, as a consequence of inaction by EPA and the Maryland Department of the Environment, the Maryland Assembly passed a bill in 1988 requiring by 1991 the implementation of advanced wastewater treatment to remove N in STPs discharging into the Patuxent.

Legislative action in 1984 also addressed the problem of nonpoint nutrient inputs. The cornerstone of this legislation was the Critical Areas Protection bill, which established a framework for managing shoreline development to minimize erosion and nonpoint-source nutrient inputs as well as for protecting critical habitats within 1,000 feet of the shore. In addition, the Sediment Control bill placed sediment control under the authority of the state; the Drainage of Agriculture Lands bill required the secretaries of Agriculture, Natural Resources, and Health and Mental Hygiene to promulgate regulations for the efficient design, construction, operation, and maintenance of agricultural drainage projects; and the Shoreline Improvement Loan bill authorized funds for projects to reduce shoreline erosion within the critical area. It should be emphasized that the implementation of measures to reduce nonpoint nutrient inputs was, and still is, a voluntary process facilitated by federal and state cost-sharing programs. Furthermore, as these actions suggest, the management of nonpoint source nutrient inputs remained dominated by the notion that soil conservation and nutrient control were synonymous, a perception that would continue through the 1980s.

Although specific actions to control nonpoint nutrient sources would not be forthcoming until the 1990s, the results of the Bay Study and continued research on nutrient runoff from agricultural lands were gaining the attention of the management community. The Chesapeake Restoration Plan, released by the Chesapeake Bay Commission in 1985, recognized the need for basin-specific nutrient control strategies and outlined implementation plans for reducing point and nonpoint nutrient inputs. These included recommendations for a ban on the use of phosphates in detergents (enacted in 1985), improved wastewater treatment throughout the state, reductions in combined sewer overflows, and the development of BMPs to reduce surface nutrient runoff (e.g., planting "buffer strips" in critical areas). The Chesapeake Bay Commission also endorsed the development and use of mathematical models to evaluate the success of nutrient control strategies "before changes can be detected physically," emphasized the importance of nonpoint sources, recommended continued research on the role of N in the Bay ecosystem, and acknowledged the publication of *Statewide Priority Watersheds for the Potential Release of Agricultural Nonpoint Phosphorus and Nitrogen* by the Maryland State Soil Conservation Committee (EPA, 1985). The latter marks an important step toward nonpoint-source nutrient control by ranking all watershed segments based on their potential for nonpoint nutrient discharge to the Bay and its tributaries.

The first significant changes to the CWA since 1972 were made in 1987. These included a change in emphasis from point to nonpoint source controls and a phaseout of construction grants for STPs by 1994 (which had provided nearly $50 billion to states for STP construction from 1972 to 1987). This shifted the burden of funding to the states and increased their authority to control toxic pollutant discharges and nonpoint sources of pollution. With the signing of the 1987 Chesapeake Bay Agreement, broad goals and priorities were established for the restoration and protection of living resources and water quality. Although the agreement contains objectives and commitments for living resources, water quality, population growth and development, public information, education and participation, public access, and governance, the centerpiece was the achievement of a 40 percent reduction in total loads of N and P to the Bay by the year 2000. This was a landmark agreement in that it established a specific and quantifiable goal that was to be reevaluated in 1991 based on the results of the monitoring program and simulation modeling.

For the first time, the agriculture community was forced to confront the question of how to reduce nonpoint source inputs from farms. With the release of *A Commitment Renewed: Restoration Progress and the Course Ahead* by the Chesapeake Bay Program Implementation Committee (EPA, 1988), it was also acknowledged that controlling the input of N would be difficult because nitrogen moves with water, in contrast to phosphorus, which moves with sediment. Finally, Maryland's Coastal Zone Management Plan (approved in 1978 and administered by the DNR), which did not explicitly address the problem of eutrophication, was modified in 1990 to include provisions for nonpoint-source nutrient control and water quality management consistent with the CWA.

The U.S. EPA Chesapeake Bay Program released its *Progress Report of the Baywide Nutrient Reduction Reevaluation* in 1992. The report presents the most accurate estimates to date of nutrient sources and loads. Results of the watershed model reaffirmed the significance of basinwide nonpoint sources (77 percent and 66 percent of N and P inputs, respectively), reinforcing the conclusion that nonpoint source N inputs must be controlled if a 40 percent reduction in inputs is to be achieved. Direct atmospheric deposition of N and P to the Bay and its tributaries was found to be relatively small, but estimates of basinwide inputs suggested that atmospheric deposition could account for as much as 35–40 percent of the total nitrogen input, a conclusion that is consistent with the findings of Fisher and Oppenheimer (1991) of the Environmental Defense Fund. Results of computer computations (using the so-called 3-D model, a three-dimensional, time-variable numerical model) suggest that 40 percent reductions in controllable nitrogen (about 20 percent of total input) and phosphorus (about 30 percent of total input) loads will increase bottom water oxygen levels by 15 to 25 percent. Based on these results and interpretations, the 1992 amendments to the Chesapeake Bay Agreement reaffirmed the commitment to a 40 percent reduction in N and P loadings by the year 2000, placed caps on these loading levels once achieved, specified basin-specific nutrient loads, and called for implementation of nutrient control strategies to achieve these loads beginning in 1993. The 1992 agreement also stipulated that the abundance and distribution of a living resource, SAV, would be used to measure the success of this nutrient control strategy.

The management of nonpoint sources of nutrients remains controversial. Although the 1992 Bay Agreement calls for reductions in atmospheric and agricultural sources of nitrogen, it is unclear how such reductions would be achieved. Both regional and nationwide reductions in nitrogen oxide emissions will be required to control atmospheric deposition. As for agricultural inputs, traditional approaches that rely on expanded implementation of BMPs (designed to reduce surface runoff and soil erosion) are unlikely to have the desired impact on N loading. The management community has interpreted recent results of watershed models and related cost-benefit analyses as indicating that the most cost-effective approaches to reducing nonpoint agricultural inputs are the control of fertilizer applications and animal waste inputs (EPA, 1992). However, the watershed model has been widely criticized as inadequate, and the quantitative effects of tuning fertilizer applications to agriculture production and avoiding accumulations of animal wastes are promising but uncertain. To the extent that groundwater pathways account for most of the nonpoint loading of N, additional measures (e.g., cover crops) that limit the movement of N into groundwater will be required.

DISCUSSION AND CONCLUSIONS

The Interplay Between Science and Management

Our analysis reveals a change in the relationship between science and management as the emphasis in nutrient control shifted from point to nonpoint sources. During the formative years, neither the science nor the management communities in Maryland perceived nutrient enrichment to be an immediate, high-priority problem (compared with thermal pollution, dredging, and the threat of oil spills). Research and management activities tended to focus on local issues and problems, a pattern that may have been reinforced by prevailing climatic conditions. Initial concerns with point source nutrient inputs coincided with a period (1962–1969) of unusually low rainfall when the problems of nutrient enrichment in lakes were first gaining national and international attention. Low rainfall and freshwater runoff have the effect of minimizing nonpoint inputs and maximizing the effects of point source inputs, which are independent of freshwater runoff for the most part. Point source nutrient inputs were targeted and the state implemented secondary treatment by constructing and upgrading STPs. These actions were driven by the federal CWAs, which provided financial incentives, and

by public and political pressures precipitated by local water quality problems that did not require sophisticated tools of science to uncover. The
relationship between point source inputs and water quality (as indicated by
such phenomena as red tides, noxious odors, and fish kills) was usually
obvious, and it was generally assumed that nutrient loading could be managed through secondary treatment to control point source loadings. Management moved out in front of the science and formulated their own "best
guess" scenarios as to the degree and kinds of nutrient reductions needed to
improve water quality.

The Potomac case may be considered an exception in this regard. Scientific research preceded management action, which appeared to be closely
coupled to new scientific information establishing quantitative relationships
between nutrient loading and water quality. Low rainfall during the 1960s
undoubtedly exacerbated conditions in the Potomac River where noxious
algal blooms, fish kills, and generally unsanitary conditions were occurring
at the doorstep of the White House. Here, secondary treatment and advanced wastewater treatment for P reversed the trend of declining water
quality, at least in the tidal freshwater reach of the estuary (Jaworski, 1990).[5]
However, the Potomac case was unique, not only in terms of the apparent
close coupling between new scientific information and management action
(which probably reflected the river's proximity to Washington, D.C., and its
role as a political showcase as much as anything else), but also in terms of
the massive expenditure of federal funds (about $1 billion) and its limited
impact on research and management in the greater Chesapeake Bay region.

A fundamental change in the relationship between science and management began to emerge with the controversy over N control in the Patuxent
River basin. The spatial displacement between the upstream location of
point source nutrient inputs and downstream effects not only set the stage
for a decade-long debate over the control of N and nonpoint source inputs,
it marked the beginning of a systemwide approach to the problem of eutrophication in the Bay and its tributaries. With this seed, the connection between
nutrient loading and water quality in the Bay as a whole began to crystalize
when research sponsored by the EPA Bay Program related widespread decline of SAV to overenrichment. At this point, science began to move out
in front of management, in part because of the complex nature of the problem and in part because of the lack of funding (including financial incentives from the federal government) to develop and implement new approaches
and technologies required to address the problems of N and controlling
nonpoint source inputs.

The management community as a whole did not acknowledge the need
to control N and nonpoint source inputs until the late 1980s when the cumulative impact of evidence from environmental research became overwhelm-
ing. With each iteration of nutrient inventories and budgets, the predomi-

nance of nonpoint sources became unequivocal. The shift from P to N limitation along the transition from fresh to salt water areas was clearly demonstrated. Studies of benthic nutrient fluxes revealed that models of water quality in the Bay would have to incorporate benthic-water column interactions into their calculations; large-scale baywide studies revealed the mechanisms by which nutrient inputs cause oxygen depletion in the main Bay, and showed that nonpoint sources were the principal cause; and current soil conservation practices were shown to have little effect on N input to the Bay. These advances could not have been made without a major research and monitoring effort by the science and management communities in the Chesapeake Bay region. Rather than depending on science information generated by research in Canadian and European lakes, the management process was increasingly guided by new scientific information on the Bay itself, information that currently influences nutrient management in estuarine systems worldwide.

Explicit actions to control N loading have been limited to point source discharges to the upper Potomac and Patuxent rivers, and strategies that target nonpoint sources of N are only now being seriously considered. On the receiving end, the 1992 Bay Agreement identifies the return of SAV as an initial measure of the effectiveness of nutrient management in the restoration of living resources and water quality. Long lags (on the order of 10 years or more) between scientific discovery and management action are a common feature of each of these cases. To some extent, this reflects a considered and informed decision-making process related to social and economic considerations and to the uncertainty of environmental science. However, the record also suggests that this is often not the case, in part because sufficient information is simply not available (increasing the uncertainty), but also because of ineffective information exchange between the science and management communities. Consequently, delays in the use of new scientific information are often related more to politics and economics (compare the histories of point source nutrient control in the Potomac and Patuxent cases) than to the quantity and quality of available information. As clearly stated by Ian Morris, the director of the Center for Environmental and Estuarine Studies of the University of Maryland (*Baltimore Sun*, July 17, 1983), "There is nothing wrong with forging ahead before knowledge of a problem is complete [because] it never is—but you need to keep close touch with good scientific study, and that close touch is being lost." A comparative study of coastal seas management in different regions from the Baltic Sea to the Inland Sea of Japan clearly shows the importance of "independent but relevant science" to the decision-making process (Morris and Bell, 1988). This study suggests that, although new scientific information rarely initiates management action, the availability of good information and scientific advice not only enhances the responsiveness and quality of management actions but also often reinforces management decisions and helps keep the management process on track.

Sources of Inertia

Inertia in the management process occurs for a variety of reasons that range from the sheer magnitude of the problem and the cost of solving it to poor problem definition, uncertainties inherent in the prediction of ecosystem behavior, and polarization between the science and management communities. Two features of the Chesapeake Bay experience that exemplify magnitude and cost stand out: the need for more STPs with advanced wastewater treatment and the need to control nonpoint sources. Clearly, reliance on a particular technology (secondary treatment) as the basis for regulating nutrient inputs has inhibited the development of alternative (less costly, more effective?) approaches and technologies (see Officer and Ryther, 1977). Furthermore, as the fiscal realities of advanced wastewater treatment for P removal became apparent in the 1970s, the Congress and the General Accounting Office became alarmed and instituted a federal "Advanced Wastewater Treatment Policy," which essentially subjected STPs contemplating advanced P or N removal to extreme scrutiny. The effect was to create a powerful disincentive for advanced wastewater treatment, especially for N. Consequently, STPs on the Patuxent did not begin to remove nitrogen until 1991, several years after cost-effective technology became available, a decade after the Patuxent Charrette, and more than two decades after scientists first began to worry about N loading to the Bay.

In the case of nonpoint sources, their diffuse nature and relationship to patterns of landuse catapulted the problem of nutrient regulation to a new level involving not only water quality and living resources but also socioeconomic forces related to population growth in the watershed. Implementation of point source controls has little direct impact on the social fabric of the population, and the costs of reducing point source inputs can be predicted with a relatively high degree of certainty based on knowledge of loading rates and the required technology. This is not the case for nonpoint inputs. Management of nonpoint sources inevitably leads to conflicts between prevailing patterns of land use (by farmers, homeowners, industry, government, etc.) and the implementation of nutrient control schemes. The cost of reducing nonpoint sources is more unpredictable because of uncertainties in loading rates and in the effectiveness of different methods of nutrient control. Thus, for justifying the social and economic costs of nutrient management, it becomes much more important to demonstrate cause-effect relationships between nonpoint sources, water quality, and the capacity of the ecosystem to support living resources. Decision makers insist on more information before implementing control measures.

Scientists and managers typically function on different time scales, resulting in tension and distrust between the two groups, which in turn inhibits effective exchange of information and consensus on problems and their solution. For the most part, environmental scientists are cast in the role of conducting research intended to further our understanding of nature. Advances occur on time scales that are dictated by factors ranging from the peer review process to the variability that characterizes populations of organisms, the ecosystems within which they function, and the climatic factors that perturb them. In contrast, managers are expected to make informed decisions and solve problems in a "timely" fashion and are often under considerable pressure to do so on political time scales that are short relative to the generation of new scientific understanding. To compound the problem, success in the science community is achieved through a process that emphasizes peer review, so there is little motivation to communicate outside the science community (except when funds are needed for research). Within the management community, success is measured, in part, by the outcome of the decision, which typically must be made before sufficient scientific information is available. The distrust that these dichotomies and lack of communication breed has two important and related consequences: (1) the management community tends to question the relevance of environmental research conducted by an independent science community, and (2) the science community tends to question the integrity of the management process.

Free from the requirement to make management decisions, scientists are much more likely to acknowledge uncertainties and the complexities of nature. For example, consider an event that occurred in 1983, the year of the first Chesapeake Bay Agreement. A headline in the *Baltimore Sun* (July 17, 1983) reads "Scientists Wary About Quick Fix for Bay." The article quotes a prominent university scientist as stating in testimony before the state Assembly that "we still don't know very much about nitrogen and the Bay. . . . The nitrogen entering the bay from farm runoff may not be as injurious, pound for pound, as that coming from . . . sewage treatment plants. . . . Buffer strips may not stop much nitrogen from running off farms . . . the bay is not purely a sink [for N, which can escape the Bay in gaseous form]." This left "decision-makers upset and confused . . . some almost cursing, 'saying what is this guy trying to do to us?'" A manager with the Maryland DNR summed up the dilemma by commenting that, "Scientists, being quite honest, present so many options that no action gets taken . . . which is our problem as managers who must take action." In a subsequent interview, Tom Horton of the *Baltimore Sun* (personal notes) quotes Secretary Eichbaum as saying "Ian's [Ian Morris] concerned about a lack of communication between scientists and us? I know he feels that way and I think he's even right, but in most of our experiences in the bay system the scientists have not provided answers as to what to do. They have sat around and complicated the issue, and that's the nature of their job. So at this stage [in the management process] it's not so unusual for scientists to recede into the background. Hopefully, the information they've given us is good enough."

The uncertainties and complexities (large data sets) inherent to environmental science have also led to, arguably, an irrational reliance on mathematical models as predictors of ecological variability and responses to anthropogenic perturbations. This tendency of the management community to view water quality models with considerable favor is understandable. Scientists who develop the models have a vested interest in seeing them used (an example of why it is important to maintain a separation of powers in terms of the generation of scientific information and its use by management). At the same time, they provide an objective means of synthesizing a great deal of information and of predicting both the causes and the consequences of eutrophication (in this case), and they take the "heat" off the decision maker (the model makes the decision). This allows the government to assess blame and institute corrective actions. Herein lies the rub. All of these are attractive (and seductive) features, but all assume that the water quality model provides an accurate representation of the real world.

The current heavy reliance on the 3-D, time-dependent, coupled hydro-dynamic water quality model to set nutrient reduction goals and evaluate the success of nutrient control programs is reminiscent of the Patuxent experience. Clearly, this model is significantly improved, but it is still an imperfect cartoon of the real world. It is so tempting to ask the model a question and then believe the answer ("mirror, mirror on the wall") when the most prudent approach is to use the model results in conjunction with other sources of information (monitoring and experimental results that reveal causation). One must also keep in mind that no single model can answer all questions. For example, the current model does not address the dynamics of littoral areas, sea grasses, or food webs. Finally, models may take many years to develop, during which time the playing field and the players may change, including expectations of what the model can and cannot do. The original intent of the model may be modest (e.g., to be used as a trial-and-error tool), but as the results are simplified again and again for nontechnical audiences, expectations can and do become unrealistic. As the cost of the model increases (in terms of time and money) and the corporate memory is lost, the model begins to take on a life of its own and the predictions become reality. Thus, there is a tendency for the management community to reach the conclusion that additional scientific information is no longer needed, a tendency that can be countered by establishing a process of periodic scientific reevaluation of the effectiveness of management actions.

Overcoming Inertia

Clearly, financial incentives in the form of federal and state funding for such actions as the construction of STPs and the implementation of BMPs have played an important role in controlling nutrient loading to surface waters. However, the authorizing legislation and subsequent appropriations are often responses to an environmental catastrophe. In a recent study of the process of environmental governance, Morris and Bell (1988) argue the case that a "major event" is required to stimulate policymakers and managers to take action on environmental issues. They suggested that, in the Chesapeake Bay region, this event was the Chesapeake Bay Study itself. Our analysis certainly supports this contention. By pulling together large numbers of scientists and decision makers from throughout the Bay and its watershed, the EPA Bay Study marked a significant departure from the course of the 1960s and 1970s. Under the auspices of the EPA, it gave rise to a governance structure that would involve citizens, government officials, and scientists in the oversight of environmental research, formulation of policy, and implementation of that policy throughout the entire Bay and its watershed (the Chesapeake Executive Council, Citizens Advisory Committee, Science and Technology Committee, and Implementation Committee). This spawned a decade of research and management activity that was unprecedented in the United States, and ushered in an era that would lead to a more systemwide perspective as the significance of nonpoint sources and water movements through drainage basins and the estuaries of the Bay became increasingly apparent.

Our analysis also suggests that Tropical Storm Agnes was an event of similar impact, which, in effect, set the stage for the EPA Bay Study. Agnes alerted a broad cross-section of the population, including scientists and managers, to the systemwide susceptibility of the Bay to inputs from land. Until Tropical Storm Agnes arrived in 1972, research tended to focus on local problems, a tendency exacerbated by the funding priorities of management agencies that emphasized the effects of power plants, oil spills, and dredging. Pritchard (July 1, 1970, *Baltimore Sun*) states that, "The emphasis on thermal pollution is obscuring the real threat to the Bay, nutrient pollution." This 200-year storm captured the attention of the entire population of the Chesapeake region, including state and federal agencies, elected officials, concerned citizens, and the scientific community. In a terrible way, Agnes reconnected millions of urban and suburban dwellers to nature. People were made keenly aware that they did not just live on a street or in a town, but also in the drainage basin of a creek, in the valley of a river. The storm dramatized how we had changed the very nature of the watershed in just a few decades, stripping the vegetation that once covered it and absorbed and slowed the runoff of rainfall, paving it for roads and parking, and roofing it over with homes. One effect of these land-use patterns was to channel the water that fell with a destructive force never seen before. In retrospect, it is clear that, although the storm delivered a "bullet to the Bay's heart," land use in the watershed had been "loading the gun and softening up the victim for many decades."

The precedent-setting 1981 Patuxent Charrette and the Chesapeake Bay Agreements that followed illustrate the importance of achieving a consensus involving a broad cross-section of a region's social fabric. Events leading up to the Patuxent Charrette and the Charrette itself underscore some of the ingredients needed to achieve a consensus on the nature of the problem and the actions that need to be implemented to solve the problem. Among the more important of these are leadership, trust, and financial incentives. A few powerful individuals had to have more than a passing interest in the problem; they needed to understand the problem well enough to justify action in the context of competing political, economic, and social forces. Such leadership was clearly demonstrated by the actions of Senator Mathias, who formulated the legislation that led to the EPA Bay Program; by Senator Fowler, whose environmental concerns led to a nutrient management plan for the Patuxent River basin; and by the state governors who had the foresight to look beyond their borders in agreeing to clean up the Bay. The Patuxent case in particular illustrates the need for trust. It is unlikely that a truly comprehensive nutrient management plan for the Patuxent River basin would have been agreed upon if it were not for a clear definition of the problem, the establishment of common goals, the existence of independent scientific advice, and mutual respect among the participating parties. In this regard, the university was viewed by Senator Fowler and his associates as a source of information from a disinterested party, an "honest broker." This was critical, as was the presence of managers within the Maryland state government who were willing to listen and even fund research that could (and did) produce evidence that the state and the EPA were wrong in insisting that N loading was not a problem (D'Elia, 1987; D'Elia et al., 1986).

The main impact of these actions and the "major" events that gave rise to them was to raise the plight of the Bay to a new level of public and political consciousness. In this context, it is important to note that, although there were (and are) few who would take exception to the course set by the 1983 Bay Agreement, important decisions were made on the basis of relatively little scientific information—decisions that would have profound social and economic consequences. Agreements were consummated by high-ranking government officials based on perceptions and the "common sense" of the day. The impact of the EPA Bay Study was not related as much to new scientific information as it was to the large number and diversity of individuals and institutions involved in the process. The real genius of the study was in synthesizing and disseminating existing environmental information and scientific understanding, and in providing the political climate needed to galvanize decision makers throughout the Bay's multistate watershed. The process itself, rather than the information it produced, led to the Bay Agreement and launched an unprecedented period of legislative, management, public, and research activity.

NOTES

1. The degradation of water quality occurs when assimilation capacity is exceeded. The degradation is expressed by such phenomena as accumulations of algal biomass, noxious algal blooms, decreases in water clarity, depletion of oxygen, and related losses of plant, animal, and insect life.

2. Environmental research is defined as activities that generate technical information about nutrient enrichment upon which the management of nutrient inputs can be based. Management is considered to be primarily a government activity that includes the formulation of environmental policy, regulations, and agreements.

3. The term *anthropogenic* is generally used to identify sources of pollutants that stem from human activities—manufacturing, farming, waste disposal, etc. For purposes of this analysis, anthropogenic nutrient inputs include inputs from point sources (such as wastewater discharges) and diffuse sources (for example, runoff from agricultural development, atmospheric deposition).

4. Hurricane Agnes caused devastating coastal flooding from Florida to New York. By the time the storm reached Chesapeake Bay, it had been downgraded to the level of a tropical storm.

5. A massive nuisance bloom of blue-green algae in the upper Potomac in 1983 was attributed to a combination of events that resulted in the release of excess phosphorus from the sediments (Jaworski, 1990).

REFERENCES

American Society of Limnology and Oceanography. 1972. Nutrients and eutrophication: the limiting-nutrient controversy. Proceedings of the 1971 Symposium on Nutrients and Eutrophication. Lawrence, Kansas: Allen Press.

Bayley, S., H. Rabin, and C. H. Southwick. 1968. Recent decline in the distribution and abundance of eurasian milfoil in Chesapeake Bay. Chesapeake Science 9:173–181.

Bayley, S., V. D. Stotts, P. F. Springer, and J. Steenis. 1978. Change in submerged aquatic macrophyte populations at the head of Chesapeake Bay. 1958–1975. Estuaries 1:171–182.

Boynton, W. R., W. M. Kemp, and C. G. Osborne. 1980. Nutrient fluxes across the sediment-water interface in the turbid zone of a coastal plain estuary. Pp. 93–109 in Estuarine Perspectives, V. S. Kennedy, ed. New York: Academic Press.

Boynton, W. R., W. M. Kemp, and C. W. Keefe. 1982. A comparative analysis of nutrients and other factors influencing estuarine phytoplankton production. Pp. 69–90 in Estuarine Comparisons, V. S. Kennedy ed. New York: Academic Press.

Bunker, S. M., and G. V. Hodge. 1982. The legal, political and scientific aspects of the Patuxent River nutrient control controversy. 8th Annual Coastal Society Conference, Baltimore, Md.

Carpenter, J. H., D. W. Pritchard, and R. C. Whaley. 1969. Observations of eutrophication and nutrient cycles in some coastal plain estuaries. Pp. 210–221 in Eutrophication: Causes, Consequences, Correctives. Proceedings of the 1967 International Symposium on Eutrophication, University of Wisconsin. Washington, D.C.: National Academy of Sciences.

Chesapeake Bay Commission. 1985. Choices for the Chesapeake: The first biennial review of the action agenda. Report to the General Assemblies of Maryland and Virginia.

Chesapeake Research Consortium. 1976. The Effects of Tropical Storm Agnes on the Chesapeake Bay Estuarine System. Publ. No. 54. Baltimore, Md.: Johns Hopkins University Press.

Chesapeake Research Consortium. 1977. Proceedings of the Bi-State Conference on the Chesapeake Bay, April 27–29, 1977. Publ. No. 61

Clark, L. J., D. K. Donnelly, and O. Villa. 1973. Nutrient enrichment and control requirements in the upper Chesapeake Bay. EPA Report No. 903/9-73-002-a.

Clark, L. J., V. Guide, and T. H. Pheiffer. 1974. Summary and conclusions: Nutrient transport and accountability in the lower Susquehanna River basin. Tech. Report 60: Annapolis Field Office, Region III, Environmental Protection Agency.

D'Elia, C. F., J. G. Sanders and W. R. Boynton. 1986. Nutrient enrichment studies in a coastal plain estuary: phytoplankton growth in large-scale continuous cultures. Canadian Journal of Fisheries and Aquatic Sciences 43:397–406.

D'Elia, C. F. 1987. Nutrient enrichment of Chesapeake Bay. Environment 29:6–33.

Eichbaum, W. 1984. The Chesapeake Bay: Major research program leads to innovative implementation. Environmental Law Reporter 14:10237–10245.

Fisher, D. C., and M. Oppenheimer. 1991. Atmospheric nitrogen deposition and the Chesapeake Bay estuary. AMBIO 20:102–108.

Flemer, D. A., D. H. Hamilton, J. A. Mihursky and C. W. Keefe. 1969. The effects of thermal loading and water quality on estuarine primary production. An interpretive report for the period August 1968 to August 1969 to the U.S. Department of the Interior, Office of Water Resources Research, Washington, D.C.

Heinle, D. R., D. H. Hamilton, and D. A. Flemer. 1970. A unified approach to research on Chesapeake Bay. Chesapeake Biological Laboratory, Ref. No. 70–32.

Heinle, D. R., C. F. D'Elia, J. L. Taft, J. S. Wilson, M. Cole-Jones, A. B. Caplins, and L. E. Cronin. 1980. Historical review of water quality and climatic data from Chesapeake Bay with emphasis on effects of enrichment. Report to EPA, CRC Publ. No. 84, UMCEES Ref. No. 80-15CBL.

Hydroscience, Inc. 1975. Wasteload Allocation Study. Maryland Department of Natural Resources, Water Resources Administration, Annapolis, Md.

Jaworski, N. A. 1981. Sources of nutrients and the scale of eutrophication problems in estuaries. Pp. 83–110 in Estuaries and Nutrients, B. J. Neilson and L. E. Cronin, eds. Totowa, N.J.: Humana Press.

Jaworski, N. A. 1990. Retrospective of the water quality issues of the upper Potomac estuary. Aquatic Science 3:11–40.

Jaworski, N. A., D. W. Lear, and J. A. Aalto. 1969. A technical assessment of current water quality conditions and factors affecting water quality in the upper Potomac estuary. U.S. Department of the Interior Tech. Rep. No. 5.

Jaworski, N. A., D. W. Lear, and J. A. Aalto. 1972. Nutrient management in the Potomac estuary. Pp. 246–273 in Nutrients and Eutrophication, G.E. Likens, ed. American Society of Limnology and Oceanography. Lawrence, Kansas: Allen Press.

Jaworski, N. A., P. M. Groffman, A. A. Keller and J. C. Prager. 1992. A wastewater nitrogen and phosphorus balance: The upper Potomac River basin. Estuaries 15:83–95.

Kemp, W. M., and W. R. Boynton. 1984. Spatial and temporal coupling of nutrient inputs to estuarine primary production: The role of particulate transport and decomposition. Bulletin of Marine Science 35:522–535.

Kemp, W. M., R. R. Twilley, J. C. Stevenson, W. R. Boynton, and J. C. Means. 1983. The decline of submerged vascular plants in upper Chesapeake Bay: Summary of results concerning possible causes. Journal of the Marine Technology Society 17:78–89.

Ketchum, E. H. 1969. Eutrophication of estuaries. Pp. 197–209 in Eutrophication: Causes, Consequences, Correctives. Proceedings of the 1967 International Symposium on Eutrophication, University of Wisconsin. Washington, D.C.: National Academy of Sciences.

Lomax, K., and J. C. Stevenson. 1981. Diffuse Source Loadings from Flat Coastal Plain Watersheds: Water Movement and Nutrient Budgets. Annapolis, Md.: Tidewater Administration, Maryland Department of Natural Resources.

Maryland Sea Grant College Program and Virginia Sea Grant College Program. 1992. Dissolved Oxygen in the Chesapeake Bay: A Scientific Consensus. College Park, Md.: Maryland Sea Grant College.

Morris, I., and W. H. Bell. 1988. Coastal seas governance: an international project for management policy on threatened coastal seas. Maryland Law Review 47:481–496.

National Research Council. 1969. Eutrophication: Causes, Consequences, Correctives. Proceedings of the 1967 International Symposium on Eutrophication, University of Wisconsin. Washington, D.C.: National Academy of Sciences.

Neilson, B. J., and L. E. Cronin, eds. 1981. Estuaries and Nutrients. Totowa, N.J.: Humana Press.

Nixon, S. W., and M. E. Q. Pilson. 1983. Nitrogen in estuaries and coastal marine ecosystems. Pp. 565–648 in Nitrogen in the Marine Environment, E. J. Carpenter and D. G. Capone, eds. New York: Academic Press.

Officer, C. B., and J. H. Ryther. 1977. Secondary sewage treatment versus ocean outfalls: An assessment. Science 197:1056–1060.

Officer, C. B., R. B. Biggs, J. L. Taft, L. E. Cronin, M. A. Tyler, and W. R. Boynton. 1984. Chesapeake Bay anoxia: Origin, development, and significance. Science 223:22–27.

Orth, R. J., and K. A. Moore. 1983. Chesapeake Bay: An unprecedented decline in submerged aquatic vegetation. Science 222:51–53.

Pritchard, D. W. 1969. Dispersion and flushing of pollutants in estuaries. Journal of the Hydraulics Division, Proceedings of the American Society of Civil Engineers 95:115–124.

Ryther, J. H., and W. M. Dunstan. 1971. Nitrogen, phosphorus, and eutrophication in the coastal marine environment. Science 171:1008–1013.

Schindler, D. W. 1974. Eutrophication and recovery in experimental lakes: Implications for lake management. Science 184:897–899.

Schindler, D. W. 1977. Evolution of phosphorus limitation in lakes. Science 195:260–262.

Seliger, H. H., J. A. Boggs, and W. H. Biggley. 1985. Catastrophic anoxia in the Chesapeake Bay in 1984. Science 228:70–73.

Smith, D. E., M. Leffler, and G. Mackiernan, eds. 1992. Oxygen Dynamics in the Chesapeake Bay. Maryland Sea Grant, College Park, Md.

Southwick, C. H., and F. W. Pine. 1975. Abundance of submerged vascular vegetation in the Rhode River from 1966–1973. Chesapeake Science 16:147–151.

Staver, K., R. Brinsfield, and J. C. Stevenson. 1987. Strategies for reducing nutrient and pesticide movement from agricultural land in the Chesapeake region. In Toxic Substances in Agricultural Water Supply and Drainage. Proceedings of National Meeting, U.S. Committee on Irrigation and Drainage, Denver, Colo.

Staver, K., R. Brinsfield, and J. C. Stevenson. 1989. The effect of best management practices on nitrogen transport into Chesapeake Bay. In Toxic Substances in Agricultural Water

Supply and Drainage: An Environmental Perspective. J.B. Summers and S.S. Anderson, eds. Denver, Colo.: U.S Committee on Irrigation and Drainage.

Stevenson, J. C., and N. M. Confer. 1978. Summary of available information on Chesapeake Bay submerged aquatic vegetation. Final Report U.S. Fish and Wildlife Service, No. 14-16-008-1255. Washington, D.C.

Stevenson, J. C., L. W. Staver, and K. Staver. 1993. Water quality associated with survival of submersed aquatic vegetation along an estuarine gradient. Estuaries 16(2): (in press).

Twilley, R. R., W. M. Kemp, K. W. Staver, and J. C. Stevenson. 1985. Nutrient enrichment of estuarine submersed vascular plant communities. I. Algal growth and effects on production of plants and associated communities. Marine Ecology—Progress Series 23:179-191.

U.S. Army Corps of Engineers. 1973. Chesapeake Bay Existing Conditions Report. Baltimore, Md.: Department of the Army.

U.S. Army Corps of Engineers. 1977. Chesapeake Bay Future Conditions Report. Baltimore, Md.: Department of the Army.

U.S. Fish and Wildlife Service. 1969. The National Estuarine Pollution Study. Washington, D.C.: Department of the Interior.

U.S. Fish and Wildlife Service. 1970. National Estuary Study. Washington, D.C.: Department of the Interior.

U.S. Environmental Protection Agency. 1982. Chesapeake Bay Program Technical Studies: A synthesis. Washington, D.C.

U.S. Environmental Protection Agency. 1983a. Chesapeake Bay Program, Choices for the Chesapeake: An Action Agenda, 1983 Chesapeake Bay Conference Report, F. H. Flanigan, ed. Annapolis, Md.

U.S. Environmental Protection Agency. 1983b. Chesapeake Bay Program: Findings and Recommendations. Washington, D.C.

U.S. Environmental Protection Agency. 1983c. Chesapeake Bay Program: A Framework for Action. Washington, D.C.

U.S. Environmental Protection Agency. 1983d. Chesapeake Bay Program: A Profile of Environmental Change. Washington, D.C.

U.S. Environmental Protection Agency. 1985. Chesapeake Executive Council, Chesapeake Bay Restoration and Protection Plan. Annapolis, Md.

U.S. Environmental Protection Agency. 1986. Chesapeake Bay Program, Scientific and Technical Advisory Committee. Nutrient Control in the Chesapeake Bay. Annapolis, Md.

U.S. Environmental Protection Agency. 1988. A Commitment Renewed: Restoration Progress and the Course Ahead Under the 1987 Bay Agreement. Chesapeake Bay Program, Implementation Committee. Annapolis, Md.

U.S. Environmental Protection Agency. 1992. Chesapeake Bay Program, Progress Report of the Baywide Nutrient Reduction Reevaluation. Annapolis, Md.

Vollenweider, R. A. 1968. Scientific fundamentals of the eutrophication of lakes and flowing waters with particular reference to nitrogen and phosphorus as factors in eutrophication. OECD Technical Report DAS/CSI 68(27), Paris, France.

Vollenweider, R. A. 1976. Advances in defining critical loading levels of phosphorus in lake eutrophication. Memorie—Istituto Italiano de Idrobilogia 33:53-83.

Walter, M. F., T. S. Steenhuis, and D. Haitch. 1979. Nonpoint source pollution control by soil and water conservation practices. Transactions—American Society of Agricultural Engineers 22:834-840.

Keeping Pace with Science and Engineering. 1993.
Pp. 39-90. Washington, DC: National Academy Press.

Tropospheric Ozone

Philip M. Roth, Stephen D. Ziman, and James D. Fine

The National Research Council (NRC) Committee on Tropospheric Ozone Formation and Measurement prefaced its recent report, *Rethinking the Ozone Problem in Urban and Regional Air Pollution*, by declaring that "ambient ozone . . . represents one of this country's most pervasive and stubborn environmental problems. Despite more than two decades of massive and costly efforts to bring this problem under control, the lack of ozone abatement progress in many areas of the country has been discouraging and perplexing." (NRC, 1991, p. vii).

Ozone (O_3) is formed in the atmosphere through photochemical reaction. The primary emitted gaseous species contributing to ozone formation are nitrogen oxides ($NO_x = NO + NO_2$) and volatile organic compounds (VOCs), including hydrocarbons and oxygenated hydrocarbons. The governing atmospheric chemistry is exceedingly complex. This complexity, which involves numerous interactions among pollutants, has hindered the development of an understanding of the most effective paths to reducing ambient ozone concentrations. Inadequate and inaccurate portrayal of emissions from both manmade and biogenic sources has also contributed to difficulties in developing successful emissions control strategies. Although more information and improved understanding is definitely needed, mitigative actions have been taken and they continue.

In this paper, we survey the regulatory framework that has been put into place over the past two decades to reduce ambient ozone concentrations, consider six pivotal issues underpinning this framework, and examine the interplay between regulation and the development of science and technology. We then attempt to extract some general conclusions that may be of value in future efforts to accommodate scientific developments in the regulatory process.

A BRIEF HISTORY OF OZONE
LEGISLATION AND REGULATION

In 1963 the Clean Air Act was enacted "to protect the Nation's air resources to promote the public health and welfare." It established "the prevention and control of air pollution at its source as the primary responsibility of the State and local governments." Federal leadership and financial assistance was requested to initiate research and development programs and to assist state and local air pollution control planning. Although the Department of Health, Education, and Welfare was authorized to execute the provisions of the act and to "recommend" air pollution control criteria, only limited enforcement authority was given to government agencies. Subsequent amendments made in the late 1960s clarified provisions such as those pertaining to federal grants for research and air pollution control programs. However, the extent to which government would regulate air quality control and planning was not delineated until the Clean Air Amendments of 1970 were adopted.

The 1970 Amendments

The 1970 amendments required the Environmental Protection Agency (EPA) to "publish proposed regulations prescribing a national ambient air quality standard (NAAQS)" that defined unhealthy concentrations of specific pollutants in ambient air (criteria pollutants). The "criteria" pollutants addressed in the NAAQS had been identified in the Air Quality Act of 1967 on the basis of existing information on the health effects of air pollutant concentrations. They included particulate matter, nitrogen dioxide, sulfur oxides, carbon monoxide, hydrocarbons, and photochemical oxidants, which were redefined as ozone when the standard was changed in 1979 from 0.08 parts per million (ppm) to 0.12 ppm. A NAAQS was later promulgated for lead. In the mid-1980s, the NAAQS for hydrocarbons was rescinded.

The 1970 amendments required that state implementation plans (SIPs) be prepared to demonstrate attainment of the NAAQSs by 1975. The plans were to be prepared by every state having one or more nonattainment areas and submitted to the EPA for approval. Should a SIP not be prepared or approved, the 1970 amendments required a federal implementation plan (FIP). The implementation plans were to focus on the reduction of criteria pollutants, and in the case of ozone, on one of its precursors, VOCs. These reductions were to be accomplished by regulating mobile and stationary sources of pollutants. Although the 1970 amendments suggested control programs to be instituted by the states in attaining the NAAQS, specific requirements for SIPs were not set out until the 1977 amendments were adopted.

As mandated by the 1970 amendments, the EPA was required to produce a list of stationary source categories to be regulated under new source performance standards (NSPSs). The NSPSs were defined by the EPA on the basis of the availability of implementable technology. New major stationary sources were defined as having the potential to emit 100 or more tons per year of a criteria air pollutant, or hydrocarbons in the case of ozone, and were required to apply the NSPSs. "Hazardous air pollutants," that is, pollutants not included in the NAAQS criteria but deemed by the EPA to cause irreversible harmful health effects, were also authorized for regulation under the 1970 amendments. Emissions of pollutants from mobile sources were to have been reduced by 90 percent from a 1970 baseline by 1975 for VOCs and carbon monoxide (CO), and by 1976 for NO_x under these amendments.

At this time, the EPA elected to institute an approach to air quality improvement that focused on managing the air resource by selectively controlling emissions rather than imposing control technology requirements on a full range of source categories. The EPA directed that plans for air quality improvement be drafted and implemented, using air quality simulation models to estimate the nature, amount, and distribution of controls needed to attain the standards. While modeling was required, models generally were inadequate or nonexistent; the gap between need and availability was sizable.

The 1977 Amendments

The 1970 amendments established a framework for federal, state, and local agencies to regulate emissions of air pollutants and called for the EPA to establish specific air quality standards. Yet, by 1975—the deadline for attainment of the NAAQS—many regions of the country still had not attained the standards. The 1977 amendments attempted to achieve attainment through stricter and more extensive control of emissions from new and existing stationary sources and through sanctions for failing to comply with provisions of the act. For example, nonattainment areas and states not preparing or implementing SIPs were potentially subject to EPA sanctions, which can include a ban on new major stationary source construction or deletion of EPA grant funds. By necessity, the 1977 amendments extended the attainment deadlines from 1975 to 1982 and included possible extensions through 1987 for ozone and CO, depending on the feasibility of attainment in some areas.

In 1977, additional legislation concerned with the permit process for major stationary sources of nonattainment pollutants was passed as part of Part D, Title I (Plan Requirements for Nonattainment Areas) of the 1977 amendments. In ozone nonattainment regions, new source review (NSR) of new major sources of VOC emissions was required to ensure that the sources employed lowest achievable emission rate (LAER) practices and offset any increases in emissions by equivalent reductions in emissions within the nonattainment area. Existing major stationary sources, which had not been previously regulated under federal mandate, were required to retrofit equipment with reasonably available control technology (RACT). RACT and NSR were applicable only to VOC, not NO_x, under the federal legislation.

Because the 1970 goal of a 90 percent reduction had not been met, the 1977 amendments established a new schedule for decreasing tailpipe emissions mobile sources. To reduce on-road vehicles' emissions that were caused by deterioration of the on-board controls, ozone nonattainment areas that applied for an extension of the attainment date to 1987 had to institute vehicle inspection and maintenance programs.

In the mid-1970s the EPA's process for reviewing the scientific basis for the NAAQSs was questioned. The 1977 amendments established the Clean Air Scientific Advisory Committee (CASAC) to review new scientific information and establish criteria on which the EPA would base changes to the NAAQSs if needed. The amendments also established a mandatory five-year review period for the NAAQSs and clarified the period during which the public would be able to comment on proposed changes to air quality standards.

By the mid-1980s it became clear that few areas would actually meet the December 31, 1987, attainment deadline for the ozone standard. The EPA's response was to propose a policy that addressed post-1987 attainment issues. The policy was based on the premise that states would be submitting new plans to demonstrate expeditious attainment. The proposal is of interest now for its content: it attempted to incorporate the latest technical and scientific information pertaining to ozone formation. It included guidance on air quality modeling and, for the first time, it addressed the potential need to reduce NO_x emissions as well as, or in lieu of, VOC emissions. However, the EPA never finalized the policy. The authority of the agency to pursue the policy was questioned, and Congress became enmeshed in debate over reauthorization of the act.

The 1990 Amendments

By 1990 some 100 areas were still classified as nonattainment for ozone, based on the fourth highest ozone concentration measured in each area for the most recent three-year period (1987–1989). This measured concentration is known as the design value. Like the 1977 amendments, the main portions of the 1990 amendments that related to nonattainment areas were directed at the ozone problem.

To define regulations commensurate with the degree of nonattainment, the 1990 amendments categorized ozone nonattainment areas as extreme, severe, serious, moderate, or marginal, based on their design values. Deadlines for attainment were set according to severity. Regulations required reasonable progress toward attainment, to be achieved through a 15 percent reduction in VOC emissions for the first six years (through 1996), followed by 3 percent per year thereafter. However, the post-1996 reduction could be satisfied by substituting a reduction of NO_x emissions for some or all of the VOC reductions after demonstrating that reducing NO_x would be as effective as reducing VOC emissions.

The definition of a "major source" was adjusted according to nonattainment categories to increase the number of sources subject to NSR and RACT regulations set forth in the act. Whereas the 1977 amendments defined a major source as one that emitted at least 100 tons of VOCs annually, the 1990 amendments lowered criteria for defining major sources in severe and serious areas to 25 and 50 tons per year of VOCs, respectively. In Los Angeles, the only area classified as extreme, a major source is any source that emits at least 10 tons of VOCs per year. In addition to an increase in the number of new sources regulated by RACT and LAER requirements, the offset requirement was increased from a ratio of 1:1 to as high as a ratio of 1:1.5 in Los Angeles.

To reduce mobile source emissions within nonattainment areas, the 1990 amendments expanded the requirements for emissions reductions to include producers of vehicle fuels as well as vehicle manufacturers. In addition to instituting an enhanced vehicle inspection and maintenance program for serious, severe, and extreme areas, the 1990 amendments mandated vapor recovery programs in fuel transfers (e.g., at gas pumps), set stricter standards for tailpipe emissions, and required a reduction in fuel vapor pressure. Alternative fuels are encouraged, and clean-burning fuels are required for fleet vehicles operating in serious, severe, or extreme ozone nonattainment areas. Reformulated gasolines with a 2 percent by weight oxygen content are required for all vehicles for the nine cities with the worst ozone problems.

The EPA's authority to impose sanctions for noncompliance with SIP criteria was modified in the 1990 amendments to include 2:1 offsets for new stationary sources. However, the agency's authority to impose construction bans was withdrawn in these amendments. The amendments also established penalties if a region failed to make reasonable progress toward attainment and included provisions to move any nonattaining area into the next higher category if it did not meet the attainment deadline.

The amendments recognized that ambient air quality problems are not restricted to consolidated metropolitan statistical areas (CMSAs), which had previously been used to define ozone air quality regions. Specifically, an 11-state transport region was created under mandate in the Northeast; this jurisdiction is to address nonattainment issues associated with the entire region. Other transport regions may also be formed by mutual agreement of the states that would be part of the region and with the concurrence of EPA.

The 1990 Title III amendments increased to 189 the number of compounds identified as air toxics and mandated their reduction through new control requirements—maximum achievable control technology (MACT). Because some of the largest emissions of the identified air toxics, such as benzene, are hydrocarbons, the MACT requirement will supplement the VOC reductions imposed by Title I. Also, NO_x reductions required by Title IV, Acid Deposition, may aid in reducing ozone.

Regulations controlling NO_x emissions from stationary sources were enacted for the first time. Coal-fired utilities, which produced 33 percent of NO_x emissions in 1989, are required to meet mandated emissions limits through the use of low-NO_x burner technology. By the year 2000 this technology should reduce emissions by two million tons from the 1980 level. This requirement, in effect, supplements those for mobile source controls specified for NO_x in Title II and for overall emissions in Title I of the 1990 amendments.

See Tables 1 and 2 for a summary of legislative and regulatory history.

California Regulation

In the 1970 amendments, all the states, except California, were barred from enacting separate mobile source regulations. California was excepted because it had historically pioneered air quality regulation and programs. The state began to address air quality issues on a local level in the 1940s when the term "smog" was first used in Los Angeles. In 1947 the California Air Pollution Control Act was passed. It established air pollution control districts within each county and empowered the districts to control emissions through a permitting process. Because the Los Angeles Air Pollution Control District was the first agency to confront the smog problem, it became a leader in controlling sources of pollutants. The district, along with the Public Health Department, attempted to introduce state control of motor vehicle emissions as early as 1958. The California Motor Vehicle Pollution Control Act created the Motor Vehicle Pollution Control Board in 1960, which was eventually replaced by the California Air Resources Board (CARB).

Since its formation, the CARB has consistently been one of the nation's most influential and innovative regulatory agencies. "The board lays down

TABLE 1 Legislative History of Control Technologies and Related Regulatory Actions

Pollutants	Source Category	1970 Amendments	1977 Amendments	1990 Amendments
Ozone		• Ambient air quality standard set for photochemical oxidants at 0.08 ppm concentration (R)	• Ambient air quality standard reviewed, reset at 0.12 ppm for ozone, 1979 (R)	• Annual percent emission reductions for VOC, and VOC plus NO_x after 1996 for O_3 nonattainment areas (L)
VOC	Mobile	• 90% reduction of tailpipe emissions from 1970 baseline for motor vehicles by 1975 (L)	• Motor vehicle tailpipe emissions reductions rescheduled (L) • Vehicle inspection and maintenance program for areas not attaining O_3 standard by 1982 (L)	• Enhanced vehicle inspection and maintenance program (L) • Motor vehicle tailpipe emissions reductions (L) • Clean fuel in fleets (L)
	Stationary	• New source performance standards (NSPS) for new major stationary sources (L)	• Existing sources subject to reasonable available control technology (RACT) (R) • New source review (NSR) for new major stationary sources to assure lowest achievable emission rate (LAER) (L)	• Not previously covered existing sources retrofit with RACT (L) • 11 new control techniques guidelines (L) • NSR for new major stationary sources to assure LAER, with NSR applicability defined by classification of area O_3 design concentration (L)

continued

46

TABLE 1 *Continued*

Pollutants	Source Category	1970 Amendments	1977 Amendments	1990 Amendments
NO_x	Mobile	• 90% reduction of tailpipe emissions from 1971 baseline for motor vehicles by 1976 (L)	• Motor vehicle tailpipe emissions reduction rescheduled (L)	• Enhanced vehicle inspection and maintenance program (L)
	Stationary			• RACT for major existing sources (L) • RACT specified for coal-fired power plants (L) • NSR for new major sources (L)

CODES: (L) = Legislative; (R) = Regulatory.

TABLE 2 Legislative History of Planning Requirements Affecting Ozone Nonattainment Areas

1970 Amendments	1977 Amendments	1990 Amendments
• State implementation plans (SIPs)	• SIP requirements for nonattainment areas • Clean Air Scientific Advisory Committee (CASAC) • Mandated 5-year reviews for NAAQS	• Redefinition of major sources in nonattainment areas • Regional transportation plans Northeast Ozone Transport Commission • Indirect source control plans

the toughest regulations, forces the biggest changes and generally blazes the path for everyone else, including the U.S. EPA," according to Matthew Wald (1992). Emission controls for nitrogen oxides were promoted by the CARB almost a decade earlier than similar efforts were undertaken by the federal government. The state ambient air quality standards for ozone and fine particulate matter (less than 10 microns in diameter—PM 10) are much more stringent than the equivalent federal standards. Requirements for additional automotive emission controls led to the development of catalytic converters, cleaner burning diesel fuels, and more efficient ignition systems in motorcycles. Other CARB programs upon which federal programs were based include reductions in VOC emissions from fuel transfer systems and solvents and propellants. The CARB has recently introduced a low-emission vehicle (LEV) program intended to dramatically reduce pollutant emissions from vehicles. The agency has also regulated emissions from small utility engines. Many of the CARB's actions have stimulated other state agencies and the federal government to consider or enact similar requirements.

PIVOTAL ISSUES IN FORMULATING REGULATIONS

This section addresses six key issues that have, or should have, motivated regulation. For each, we describe the evolution of understanding over the past two decades, discuss how the issue has been addressed in the regulatory process, and examine the extent to which available knowledge has been reflected in regulation.

Formulation of the Standard

• What is an appropriate air quality standard for ozone, in terms of concentration and averaging time?

Setting the Original Standard

The 1970 amendments to the Clean Air Act mandated that the EPA set primary and secondary air quality standards for the concentrations of oxidants in ambient air. The primary standard was to serve as a regulatory reference for defining acceptably clean air. Oxidants and other pollutant standards were defined to be those which "in the judgment of the [EPA] Administrator, . . . allowing for an adequate margin of safety, are requisite to protect human health." (The secondary standard, which we will not address here, is "requisite to protect the public welfare." The prevailing primary and secondary standards for ozone are quantitatively equal.) Because of the intrinsic uncertainties associated with the definition of criteria used in establishing the standard, the dearth of substantiating scientific data, and the health and economic implications of the standard, the standard setting process has been mired in controversy from the outset.

The criteria used to establish the standard were difficult to define. The term "threshold" referred to the oxidant concentration at which exposure results in an "adverse health effect." Yet, the degree and type of health effects occurring in clinical tests vary depending on the subject (*Federal Register* [FR], Vol. 44, February 8, 1979), resulting in a range of ozone concentrations at which adverse health effects occur, rather than a threshold (Landy et al., 1990). The EPA noted this fundamental point of confusion: "the adverse health effect threshold concentration cannot be identified with certainty" (*FR*, Vol. 44, February 8, 1979). Moreover, the concept of a threshold implies the level of a measure at which no health effect occurs, which, by virtue of this property, is inherently difficult to determine. As a consequence of this dilemma, establishing the human response that barely constitutes an "adverse health effect" was less a scientific observation than a policy decision (Landy et al., 1990). In short, identification of a threshold appears to be an unavoidably uncertain determination.

Some believe that it is appropriate to identify and protect the most sensitive population group when establishing a standard because reactions to oxidants depend on the sensitivity of an individual. In 1971 the focus was on asthmatics (*FR*, Vol. 36, April 30, 1971), but subsequent research has suggested that other groups, such as children and the elderly, may be more sensitive because of physiological characteristics or exposure frequency (Lippmann, 1989).

Adding to the uncertainty were the criteria for a "margin of safety." An adequate margin could not be defined on the basis of scientific data. Rather, the EPA had to make a value judgment (Landy et al., 1990). In addition, the scientific community did not have a complete understanding of the significance of long- versus short-term exposure to air pollutants (Lippmann, 1989). Thus, it was difficult to define a time increment for measuring oxidant concentrations—the "averaging time." Defining an averaging time, a margin of safety, the most sensitive group, adverse health effects, and an oxidant threshold concentration required making assumptions founded in uncertainty.

In 1970 and 1971 the EPA conducted an intensive review of the health effects literature. Virtually all the studies reviewed failed to provide definitive information on the health effects of ozone at low concentrations. After much consideration, the EPA determined that a study conducted by Schoettlin and Landau (1961) provided the most acceptable scientific basis for setting an oxidant standard (*FR*, Vol. 36, April 30, 1971). This study reported an increased incidence of asthmatic attacks on days when the ozone concentration exceeded 0.10 ppm. Obviously, a single study could not provide sufficient information upon which to base an ozone standard. The EPA was aware of this but was nevertheless required to set a standard by the 1970 CAAAs. The paucity of available information meant that the EPA had to make a policy decision concerning the threshold at which health effects occur (Landy et al., 1990). When the EPA selected a standard of 0.08 parts per million of photochemical oxidant (later changed to ozone) in ambient air measured over a 1-hour period, a debate ensued over its adequacy and appropriateness. The controversy reflected concerns about uncertainties in defining the standard, skepticism about its scientific basis, and the health and economic implications of the policy decision.

Although the EPA wrote, "the Clean Air Act does not permit any factors other than health to be taken into account in setting the primary standards" (*FR*, Vol. 36, April 30, 1971), special interest groups aligned themselves on either side of the issue. Those likely to be burdened with the cost of attainment—industry and municipalities faced with air quality problems— were critical of what they viewed as an overly protective standard. Environmental groups supported the 0.08 ppm standard, or an even tighter standard. When a review of the Schoettlin and Landau study, conducted at the request of industry, revealed suspicions about the results, the scientific basis for the standard became questionable (Landy et al., 1990).

Revising the Standard

The need for a formal scientific review to support the EPA was acknowledged by legislation in the 1977 CAAAs. The Clean Air Scientific Advisory Committee was established to review new scientific findings for inclusion in a "criteria document," which was intended to centralize all current information related to research on the health effects of ozone and to provide the CASAC's recommendation regarding an appropriate threshold. (Criteria documents had been prepared before 1977 as well.) It was to be the document upon which the EPA would base standards. In addition to the CASAC's input, the 1977 CAAAs mandated a public comment period and mandatory review of the standard every five years.

Scientific information on the health effects of ozone, based on clinical tests on humans and animals and on epidemiological studies, was expanded during the mid-1970s (see Landy et al., 1990). Four clinical studies (Delucia and Adams, 1977; Hackney, 1975; Linn, 1978; and von Nieding, 1977) provided contradictory results. Two suggested that health effects occurred at ozone concentrations of 0.15 ppm in healthy young men, whereas the remaining two showed no effects in asthmatics and young men at levels of 0.20 ppm and 0.25 ppm, respectively. In reviewing the standard-setting process, Melnick (1983) observed that the Delucia and Adams study was "the single most important clinical evidence relied upon by the EPA" when recommending a revised standard. While those advocating a more relaxed standard argued that studies showing health effects at ozone concentration levels of 0.20 ppm and 0.25 ppm should be given equal consideration, EPA administrators were more concerned with studies showing health effects at low levels of ozone concentration because they felt they had a legal obligation to set precautionary standards (Melnick, 1983).

In addition to clinical studies, at least seven epidemiological studies were conducted in Japan and the United States. These results were suspect, however, because of the lack of controls employed in the experiments and the difficulty of applying the results to setting ozone thresholds. Also, while several animal studies had been conducted, there was no method for relating these results to human health.

Scientific information in the late 1970s, though more expansive, was neither complete nor definitive. Most studies were inappropriate for supporting policy decisions because of ambiguous results (Landy et al., 1990). Consequently, the CASAC was faced with uncertainties similar to those that the EPA had faced earlier in defining criteria for the standards. Defining the most sensitive group, "adverse health effects," a margin of safety, and the threshold value continued to pose problems, as they had during standard-setting carried out in 1971. The role of the CASAC was further complicated by a renewed emphasis on the economic feasibility of implementing the standard. Although apparently dismissed in the 1970 CAAAs, economic concerns resurfaced in the political climate of the late 1970s. President Carter stressed cost-benefit analysis in justifying government regulation, and he created the Regulatory Analysis Review Group (RARG) to analyze the economic impact of government regulations (Landy et al., 1990).

At the same time, the Science Advisory Board (SAB), of which the CASAC is a part, was asked by the EPA administrator to review the criteria document. In fact, the board served as an important critic of all three drafts of the criteria document presented by the EPA. Also, the Advisory Panel on Health Effects of Photochemical Oxidants (known as the "Shy Panel") and a team of decision science analysts were convened by the Office of Air Quality Planning and Standards to recommend alternative levels of a standard, with rationales, to the EPA administrator (Landy et al., 1990; Melnick, 1983). To critique the criteria document constructively, each group had to address and resolve uncertainties associated with the information used in setting the standard. (The EPA, the SAB, and thus the CASAC were themselves barred from considering economics in their deliberations.) While the Shy Panel and decision science analysts recommended a standard of 0.08 and 0.15 ppm, respectively, the SAB advocated relaxing the standard to the degree supported by industry. Ultimately, the EPA did not incorporate the opinion of the SAB into its suggested ozone standard.

In 1978 the EPA's Office of Research and Development (ORD) prepared a criteria document in which health effects were judged to occur potentially in sensitive persons at exposure levels of 0.10 ppm (Landy et al., 1990). Although the document recognized that the "lowest observable adverse effects levels" (LOAEL) had not been observed in any clinical studies on humans at exposure levels below 0.15 ppm ozone, it noted experiments showing respiratory infection in animals, and possible interactive effects involving other air pollutants. The OAQPS's Shy Panel and risk assessment analyses supported the 0.15 ppm ozone concentration as the threshold point for health effects. However, the SAB never approved this version of the criteria document because its members believed the conclusions were based too heavily on flawed studies indicating health effects below 0.25 ppm ozone concentration, such as the Delucia and Adams study (Melnick, 1983).

Soon after the release of the criteria document, the EPA recommended that the standard be relaxed to 0.10 ppm. At the ensuing series of public hearings, the American Petroleum Institute (API) suggested a 0.25 ppm ozone standard based on the "lack of firm evidence of significant adverse health effects near the proposed standard." In contrast, the American Lung Association and Environmental Defense Fund argued that the exposure to concentrations of 0.08 ppm ozone had not been proven safe. In addition, concerns about the cost of implementing an overly conservative standard led RARG to recommend an ozone concentration standard of 0.14 ppm. (See Landy et al., 1990, for a summary of the various responses.) Clearly, as Melnick (1983) summarized, "the answer to the question of where health effects begin usually depends on whom you ask."

The EPA administrator eventually selected a "compromise" standard of 0.12 ppm incorporating a 1-hour averaging time. According to Melnick (1983), this represented a health effects threshold of 0.15 plus a 20 percent margin of safety. Although the criteria document identified options for the EPA administrator to consider, including a more conservative formulation, the EPA administrator had to consider public policy issues and feasibility of implementation. Considering the uncertainties surrounding definition of the standard and the incompleteness of scientific information, the administrator could conclusively identify only a range in concentration rather than a specific concentration level. In the end, as Douglas Costle, the administrator, later stated, selection of the concentration level "was a value judgement" (Landy et al., 1990). The role of scientific information in the selection process was further put into perspective when, during the 1988 review of the standard, the EPA wrote:

> Although scientific literature supports the conclusion that particular ozone concentrations and exposure patterns may pose risks to human health, scientific data can only identify the limits of a range within which a standard should be set. Specific numeric standard levels, frequency of allowable exceedances, and averaging times are largely a public policy judgement (EPA, 1988, p. viii-8).

Subsequent Review of the Standard

Numerous new studies have been published that provide additional scientific data related to setting ozone concentration standards. Lippmann (1989) cites 110 studies published between 1979 and 1988 in "Health Effects of Ozone: A Critical Review," observing that "we . . . know a great deal about some of the health effects of ozone." Still, uncertainty is an important factor in selection of the ozone standard. The concept of a threshold is intrinsically difficult to define, as admitted by both the EPA and the scientific community (Landy et al., 1990). In addition, an adequate margin of safety has not yet been established. The CASAC did reach consensus on the definition of adverse health effects, which are described in Lippmann (1989). Sensitive population groups are also further defined; they include asthmatics and people who regularly exercise outside.

In 1988 only half of the members of the CASAC believed the current ozone standard was adequate to protect human health (Lippmann, 1989). The significant existing concerns relate to averaging time. Until recently, concerns over averaging time have been usurped by the larger issue of ambient air concentration thresholds. However, recent scientific data indicate that the effects of long-term exposure to low concentrations of ozone may have greater adverse health effects than short-term exposure at peak concentrations (Lippmann, 1989). Effects from exposure to ozone have also been shown to become progressively more significant as the duration of exposure increases (Lippmann, 1991). The need to develop further scientific data on these potential health effects has been pointed out by the CASAC. The EPA confirmed the need for additional research when, in its summary report on the 1988 review of the ozone standards, it stated,

> the review of the need for longer-term [ozone] standards by EPA should be continued. Because there is a good data base available on 1-2 hour exposures, the staff recommends that review of this scientific information be closed out. With this portion of the review complete . . . the [EPA] Administrator will be in a position to make a regulatory decision on how and when to best act on the 1-hour standard (EPA, 1988, p. xi-6).

Gathering additional scientific information on the long- and short-term health effects of ozone will have a twofold result. First, the need to revise or expand the 1-hour averaging time standard may be confirmed. Second, should such LOAEL be observed in longer averaging times, the ozone standard may need to be tightened. As Lippmann observed after reviewing new scientific information in 1991:

> [b]ecause the various transient effects on lung function are more directly proportional to cumulative daily exposure than to peak hourly concentration, . . . the degree of protection provided by the current [standards] is much lower than previously believed (Lippmann, 1991, p. 1956).

As noted earlier, questions about the costs of attaining the present standard arose during the standard revision process in 1978. However, it is difficult to evaluate the merits of incurring the costs of attainment unless the savings are compared with the associated benefits. Until recently, there were few attempts to quantify both sides of this complex equation, partly because methodologies for doing so were unavailable. Recently, two studies have been published that attempt to make this comparison. Krupnick and Portney (1991) examined the costs associated with attaining the present standard in the Los Angeles Basin and concluded that they significantly outweighed the benefits. However, the authors were careful to note that uncertainties were rife in their estimates and that the study actually was intended to stimulate discussion:

> Finally, implicit in our discussion is discomfort with the premises on which our nation's air quality standards are now based. If, as seems likely, there are no pollution concentrations at which safety can be assured, the real question in ambient standard setting is the amount of risk that we are willing to accept. This decision must be informed by economics. Although such economic considerations should never be allowed to dominate air pollution control decisions, it is inappropriate and unwise to exclude them (Krupnick and Portney, 1991, p. 527).

Hall et al. (1992) used a methodology developed under contract to the South Coast Air Quality Management District to make similar comparisons for ozone (as well as an assessment of PM-10) and estimated that the benefits of attaining the ozone standard would be $1.2 billion to $5.8 billion annually, with a best estimate of $2.7 billion. In presenting the implications of the study, the authors discuss the state of the methodology relative to policy decision making.

> One (implication) is that benefit estimation has not reached the maturity that policy-makers would like and cannot yet provide definitive answers to difficult economic questions (Hall et al., 1992, p. 816).

As Lippmann noted in the conclusion of his 1991 review, in reference to the need for a more definitive data base on the chronic effects of human exposure to ambient ozone:

> Further controls on exposure to ambient ozone will be extraordinarily expensive, and will need to be very well justified . . . . It is therefore important that health scientists and control agency personnel understand the nature and extent of human exposure and the effects they produce to communicate health risks effectively to the public, and to help develop realistic priorities and feasible options for reducing human exposures (Lippmann, 1991, p. 1956).

In 1990, Congress decided to preserve the present standard-setting process rather than to adopt an approach that explicitly takes into account both benefits and costs. However, some (e.g., Krupnick and Portney, 1991) argue that costs should be explicitly examined in the process of developing legislation. If this is to be done, Congress must confront two issues. First, methodologies for estimating benefits by assessing the cost savings of reduced risks to health are likely to remain imprecise for some time. Second, and more important, a benefits-cost approach to valuing human life is likely to be perceived negatively by the public. In the past, Congress has been very reluctant to deal directly with the issue of "value of human life," and we should not expect a major shift now.

Finally, as evidenced by the discussion of earlier efforts, the standard-setting process transcends scientific knowledge alone. The EPA administrator must also take into account a number of factors that affect the contemplated legislation or are put into effect by it—for example, the legal and sociological implications of a proposed change. The standard-setting process is not well defined. Decision making involves weighing many factors unique to the situation at hand; thus, the process does not lend itself to specification a priori.

The standard-setting process and its attendant problems illustrate the difficulties of formulating environmental legislation that includes unambiguous information on anticipated effects or possible outcomes in circumstances that are characterized by substantial uncertainties. The current state of knowledge seems never to be adequate; additional information is always needed. Nevertheless, decisions must be made. Science provides the clearest available statements· of current knowledge, including uncertainties and information gaps. This information is incorporated into legislation after policy deliberations that take into account socioeconomic factors and values, as well as knowledge.

Emission Control Strategies

• Should reductions in VOC or NO_x emissions or both be favored in pursuing attainment of the ozone standard?

Both VOCs and nitrogen oxides are emitted pollutants (actually, sets of pollutants) that participate in the atmospheric chemical reactions leading to the formation of ozone. For more than 20 years, VOCs have been the primary targets for emissions reduction. At various times during this period, however, the question of whether to focus on VOC reductions, NO_x reductions, or both has been the subject of discussion and debate. The issue once again is receiving significant attention.

Two main bodies of data—the results of smog chamber experiments and examination of data collected at monitoring sites in several U.S. cities—led officials of the National Air Pollution Control Administration (NAPCA) to conclude in 1970 that reduction of VOC emissions would be the appropriate mechanism for effecting reductions in ambient ozone concentrations. Los Angeles officials believed that NO_x controls should be implemented as well, on the basis of smog chamber studies carried out locally. NAPCA scientists questioned the reliability of these results and tended to dismiss their significance. Both the monitoring data and the federal chamber work were limited in extent and in accuracy because measurement methods and chamber techniques were still in evolution. Yet they constituted the primary available evidence (personal communication with B. Dimitriades, 1992).

Other factors influenced decision making. Eye irritation experienced in the Los Angeles Basin was attributed to the presence of formaldehyde, acrolein, and possibly other organic compounds, and controlling VOCs would mitigate this insult. Cost analyses indicated that reducing VOCs would be less expensive, per ton of emissions, than reducing NO_x. Methods that appeared effective in reducing VOC emissions from motor vehicles also appeared effective in reducing carbon monoxide emissions, another pollutant of concern. The net result was a determination that reduction of VOCs was the appropriate measure for reducing ambient ozone concentrations (personal communication with B. Dimitriades, 1992).

At an early stage in federal efforts to reduce precursor emissions, an ambiguity arose in control philosophy. The 1970 and 1977 CAAAs required a 90 percent reduction in VOC and NO_x emissions from motor vehicles. Legislators presumably saw "cleanup" of automotive emissions as an opportunity to challenge the automotive industry to reduce both precursors by developing novel control technology. However, this focus on controlling NO_x as well as VOCs for motor vehicles did not carry over to stationary sources for some time: federal requirements for regulating NO_x emissions from these sources were not instituted until 1990. (To be sure, NO_x emissions from stationary sources often affect ozone control strategies quite differently than NO_x emissions from mobile sources and surface sources in general. However, this issue seems not to have been explicitly addressed either.)

In the mid-1970s the Bureau of Mines presented findings of smog chamber experiments in which auto exhaust was irradiated (Dimitriades, 1972). The results suggested that, under conditions typifying ambient concentrations in urban areas, control of NO_x emissions may lead to increases in ozone concentrations, or at least reductions in the size of the decrease that would be achieved through control of VOCs alone. Although the EPA accepted these findings as evidence supporting its earlier policy decisions, criticism of this position followed. Primary concern focused on the tenuous technical basis for the so-called Appendix J curve (see Figure 1)—a plot of ozone versus

Maximum Measured 1-hour Photochemical Oxidant Concentration (ppm)

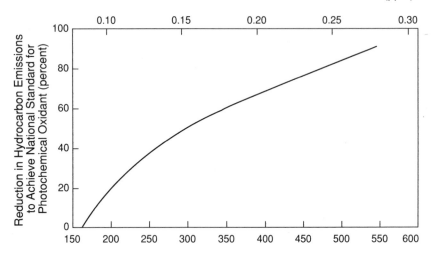

Maximum Measured 1-hour Photochemical Oxidant Concentration (μg/m³)

FIGURE 1   "Appendix J" curve. Required hydrocarbon emission control as a function of photochemical oxidant concentration. NOTE: No hydrocarbon or photochemical oxidant background assumed. SOURCE: Environmental Protection Agency (1971).

VOC concentrations for cities where such data were available (*FR*, Vol. 36, April 30, 1971). Each point plotted represented the peak ozone and VOC concentrations for a city. The "Appendix J" curve was a nonlinear envelope that confined all plotted points below it; it thus was intended to represent an upper bound on feasible combinations of maximum VOC and ozone concentrations. The plot suffered from several deficiencies. Data were available for only six cities. NO_x data were virtually nonexistent; the data that were available were of questionable accuracy. (The VOC data were inaccurate as well.) Critics were concerned that conclusions were being drawn from inadequate evidence—few data and weak analytical relationships. In short, Appendix J was viewed as being "too empirical" and very likely unreliable. The EPA's response to this criticism was the "son of Appendix J," the conceptual version of what was very soon to become the EKMA—the empirical kinetic modeling approach (Dimitriades, 1977).

During the last half of the 1970s, considerable effort was devoted to developing information about chemical reaction rates and product distributions and to studying collective chemical dynamics in smog chambers. This information was used to estimate parameters for mathematical representations of atmospheric chemical dynamics (chemical mechanisms) that were under intense development (Atkinson and Lloyd, 1984). The data were also used to evaluate the capabilities of mechanisms to emulate chamber observations. These efforts produced substantially improved chemical mechanisms that were used in the EKMA and in photochemical models.

The development of models capable of simulating the dynamics of atmospheric processes that lead to ozone formation (commonly termed photochemical models, or "grid" models) began about 1970. By 1973 the EPA had committed its research efforts to supporting continued development of the Urban Airshed Model (UAM) (Reynolds et al., 1973). The UAM, modified a number of times, is the model currently designated for use by the agency (EPA, 1991). The model saw very limited application—solely in the Los Angeles Basin—during the mid-1970s; more widespread and intense application began in 1979 and has continued to this day.

Because the UAM was designed to simulate both meteorological and chemical processes and because it is both spatially and temporally resolved, it was judged a potentially more reliable simulator of atmospheric dynamics and ozone formation than its predecessors. By 1982 both the EKMA and UAM were recommended for use in specific applications; by 1986 the UAM use was generally favored whenever data were available to support the application.

The EKMA models produce two-dimensional plots of VOC and NO_x emissions, on which are drawn contours of constant peak ozone concentrations (see Figure 2). The shapes of these contours at lower VOC and higher NO_x emissions levels (see, for example, point A) indicate that decreases in VOC emissions will reduce peak ozone concentrations, but decreases in NO_x emissions will have the opposite result. Urban Airshed Model simulations, using the same chemical mechanisms and taking into account full three-dimensional flow patterns as well, produced similar findings. Since a number of urban areas appeared to have "operating points" near or above "the ridge" in the diagram, EKMA modeling carried out from 1977 through 1982 generally supported VOC controls. However, the scientific community was aware of the uncertainties and limitations in modeling, of the limitations in aerometric data bases, and of the uncertainties in emissions representations: the evidence was incomplete.

During the historical period under review, evidence was offered and concerns were expressed at various times about the potential merits of reducing NO_x emissions. Results of chamber studies conducted by the Bureau of Mines in the mid-1970s indicated that at high ambient VOC/NO_x ratios, it would be more beneficial to reduce NO_x emissions (Dimitriades, 1977). The EPA believed that these conditions prevailed in rural, not urban, areas. The agency's concern was the reduction of ambient ozone in population centers; VOC controls were expected to suffice in these critical areas.

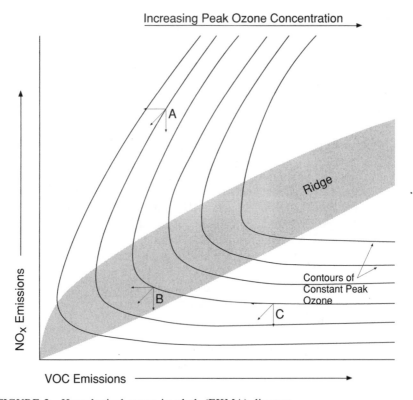

FIGURE 2   Hypothetical ozone isopleth (EKMA) diagram.

Once "EKMA diagrams" were generally accepted as being schemati-
cally representative of the VOC-NO$_x$-ozone system, they were used to demonstrate the potential merits of NO$_x$ control in areas where the "operating
point" characteristic of the area was located "below the ridge," that is, at a
location on the diagram (see Figure 2), such as point C. The problem lay in
demonstrating that a particular urban area was characterized by a point
"below the ridge." For most urban areas, data needed to support such a
contention were either inadequate or lacking.

Another general characteristic of the EKMA diagram for most urban
areas is the more favorable "cost-benefit ratio" of implementing either NO$_x$
or VOC control alone, compared with that achieved by controlling both.
The shape of the isopleths of constant ozone dictate that moving from an
"operating point" (such as point A or C) toward the origin (combined control) is always less efficient (or more costly) than moving parallel to either
axis (control of one precursor or the other). Moreover, greater control of

VOC is needed to achieve a specified ozone concentration when NO_x is reduced than when it is not. Thus, if air pollution in an urban area is of concern, the diagram suggests control of either, but not both, precursors. However, if there is reason to reduce emissions to mitigate another air pollution problem, such as PM-10, the *overall* effectiveness of a strategy aimed at meeting two specified targets must be assessed; in this case, control of both precursors may be preferred. The issue of overall effectiveness also arises in considering the transport of pollutants as well; "hybrid" control strategies may be needed to meet both local and regional air quality goals.

Evidence of a different nature supporting the potential benefits of NO_x control in certain circumstances emerged during the mid-1980s. Trainer et al. (1987) found, through observations made during periods of elevated ozone concentrations in rural locations in Colorado, that NO_x concentrations were very low. In such circumstances, control of VOC was likely to be ineffective; NO_x control would be required in order to reduce ozone appreciably. In carrying out EKMA modeling for Atlanta, Chameides et al. (1988) found that when biogenic emissions were included in the inventory, the magnitude of VOC reductions needed to attain the ozone standard increased to a level surpassing the estimated NO_x control requirement. In effect, the high biogenic emissions in the Southeast appear to elevate the ambient VOC/NO_x to levels favoring NO_x control over VOC control.

At about this time, the congressional Office of Technology Assessment (OTA) issued a report entitled, *Catching Our Breath* (OTA, 1989) that questioned whether VOC control was sufficient to ensure attainment of the ozone standard. From a series of analyses examining air quality in a large number of U.S. cities, the OTA concluded that

> [L]ocal controls on VOC emissions cannot completely solve the Nation's ozone problem. New control methods will be needed, but looking beyond the traditional controls raises challenging new technical and political issues. One promising approach for some areas is controlling NO_x, both locally and in areas upwind of certain nonattainment cities.

In addition, the report stated that

> Congress might wish to require studies to determine which areas would indeed benefit from NO_x controls. On the other hand, it may instead wish to require controls everywhere, but allow for exemptions in places where they are useless or counterproductive in reducing ozone (OTA, 1989, p. 20).

During the period from 1989 to 1992, a series of disparate studies—monitoring, modeling, and data analyses—strongly indicated that inventories in California substantially understate overall VOC emissions, perhaps by a factor of two. These studies included measurements of VOC, NO_x, and CO in the Van Nuys tunnel and comparison with emissions estimates (Ingalls et al., 1989), remote instantaneous roadside measurements of emissions from individual vehicles and analyses of data acquired during inspection of automobiles in California (Lawson et al., 1990), analyses of ambient and emissions ratios (Fujita et al., 1992), UAM simulations carried out by CARB staff (Wagner and Wheeler, 1991), and photochemical air quality model simulations (Harley et al., 1992). The studies showed that understatement of VOC emissions appears to be in large part attributable to biases in estimated mobile source emissions. If, indeed, total VOC emissions have been underestimated by a factor of two or thereabouts (and boundary and initial conditions are maintained approximately the same), then correcting them would increase the emitted VOC/NO_x by a similar factor, shifting the reactive mixture represented in models toward an NO_x-lean environment and shifting relative emissions control benefits from VOC toward NO_x. (In fact, either boundary conditions or initial conditions or both have probably been increased in modeling exercises within their ranges of uncertainty to compensate for the unrecognized underestimation of VOC emissions. Thus, one must determine the aggregate loading of emissions and initial and boundary conditions to establish the extent to which the estimated relative benefits of VOC versus NO_x controls shift in correcting for underestimated VOC emissions.)

Continuing work during the 1980s confirmed the inhibitory effect of NO_x emissions controls in subregions near major source centers. Such findings merely heightened the dilemma: NO_x controls appear to be beneficial in some conditions, detrimental in others. Currently, photochemical modeling (i.e., "the modeling system"—a meteorological model, an air quality model, an emissions representation, and requisite supporting data) simply is not sufficiently reliable to establish unambiguously the circumstances and extent of benefit and detriment. Moreover, sufficient data are not available for most areas to make a reliable case through analysis. Thus, the debate continues.

In densely populated, multiple-city regions such as the Northeast, precursors and ozone can be transported from one metropolitan area to another, influencing pollutant formation in the downwind area. This situation is exceedingly complex and requires considerable study to gain a proper understanding of control requirements. Work to date that has indicated the benefits of employing NO_x controls for a range of circumstances is intriguing. However, these results cannot be viewed as definitive because of the inadequacy of supporting data bases and certain restrictive assumptions associated with the Regional Ozone Model (Reynolds et al., 1992).

In 1990 California enacted legislation that, if implemented, would mandate new standards for vehicle emissions and regulate all types of fuels equitably; permissible emissions limits are related to the "reactivity" of the fuel. California regulation is based on a reactivity adjustment factor (RAF), where RAF is equal to the ratio of (1) the amount of ozone formed per gram of tailpipe emissions of alternative fuel to (2) the amount of ozone formed per gram of tailpipe emissions of conventional gasoline in a comparable vehicle.

While considerable attention is being given to developing reactivity factors in California, research being carried out at the University of North Carolina (Jeffries and Crouse, 1992) suggests that, in environments where the pool of free radical species is sufficient to promote and sustain reactions, about half of the total ambient ozone is formed through the oxidation of the least reactive species, notably methane, carbon monoxide, and paraffinic hydrocarbons. At the same time, reactive, radical-forming species, such as formaldehyde, are needed to ensure the sufficiency of the radical pool. Results of continuing research may clarify the relative importance of reactivity in determining the contributions of VOC constituents to ozone formation. Should research findings indicate that less, as well as more reactive VOCs contribute significantly to ozone formation, this would suggest that it may be difficult to control adequately all sources of organic compounds because of their often high concentrations, ubiquitousness, and sometimes natural origins. An implied consequence of such an outcome is the need for NO_x controls.

In 1991 the National Research Council issued *Rethinking the Ozone Problem in Urban and Regional Air Pollution.* In that report (p. 11), the authors state that "to substantially reduce ozone concentrations in many urban, suburban, and rural areas of the United States, the control of NO_x emissions will probably be necessary in addition to, or instead of, the control of VOCs." Considering the findings of past studies, areas likely to benefit from NO_x control include rural areas (very low NO_x concentrations), regions having high levels of biogenic emissions (such as the Southeast), and cities where industrial emissions of VOCs are high. In addition, NO_x controls may be beneficial in urban areas that are located in a multicity "high ozone" region that experiences long distance transport. However, atmospheric dynamics and emissions patterns, including biogenics, at the regional scale merit considerable study before conclusions are reached. The NRC report has rekindled and refueled the VOC versus NO_x debate; the issue is stimulating vigorous inquiry.

This discussion points out the lack of clarity that has plagued the VOC versus NO_x controversy for nearly two decades. Although the debate resurges from time to time, the EPA has basically promoted VOC control as the principal path to attaining the ozone standard, with two main exceptions. First, as previously noted, in both the 1970 and 1977 amendments Congress mandated 90 percent reductions in NO_x, as well as VOCs from motor vehicles. Second, in Title I of the 1990 amendments, Congress mandated 15 percent reductions in VOCs over the first six-year period and 3 percent reductions per year of some combination of VOC and NO_x thereafter until the area attains the ozone standard. During the post-1996 period, NO_x reductions can be substituted for some or all of the VOCs. Furthermore, Title I, Section 182(f), requires implementation of NO_x controls, RACT, and NSR for stationary sources for urban areas classified as extreme, severe, serious, and moderate for ozone-related air pollution, unless the administrator determines that some or all reductions in NO_x would not contribute to attainment or would not result in either net air quality or ozone benefits. (Note that the possibility of exempting NO_x does not extend to VOCs. Should reductions in VOC emissions prove to be not beneficial for an area, the reductions must still be made. Thus, the 1990 amendments reflect a bias toward VOC control.) Congress also required a reduction of two million tons per year in NO_x emissions from fossil-fuel-fired power generating stations. Although this Title IV requirement is motivated by a desire to reduce the deposition of nitric acid, it also reduces the amount of NO_x available for participating in ozone formation.

Studies conducted or reported beginning in the mid-1980s appear to have had the greatest influence in promoting the inclusion of NO_x emissions reductions in the 1990 amendments. Until that time, evidence supporting NO_x controls appears to have been viewed as either weak or unreliable. Thus, NO_x emissions from stationary sources remained unregulated at the national level for two decades. (NO_x emissions from stationary sources have been regulated in California for over a decade through a long series of rulemakings. CARB pursued this alternative course in the belief that attainment of the standard could not be achieved in the state without taking such actions.) However, as of this writing, the merits of NO_x controls are not well established. For example, Wagner et al. (1992) found, through a comprehensive series of sensitivity runs, that NO_x controls do not appear to contribute to reductions in ozone concentrations and population exposure in the South Coast (Los Angeles) Air Basin.

In summary, available evidence suggests that there is no current justification for regarding either VOC or NO_x control as being *generally* more effective. At the same time, the traditional case-by-case approach that prevailed prior to 1990 has been replaced by one in which presumptive controls are mandated, but the NO_x component can be altered in accordance with area-specific modeling results and other related evidence.

Range of Objectives

• Should controls of VOCs and NO_x be based on a desire to reduce PM-10, visibility impairment, and acidity as well as ozone?

• Should planning for improving air quality, focusing on all elements, be integrated? If so, is integration feasible?

Volatile organic compounds, NO_x, and sulfur oxides ($SO_x = SO_2 + SO_3$) are precursors to the formation of suspended particles (PM-10 and PM-2.5), atmospheric acidity and acidic deposition, and impairment of visibility, as well as to ozone and other atmospheric oxidants. Nitrogen oxides and other pollutants influence free radical concentrations and thus atmospheric chemistry in general, as well as the concentrations of free tropospheric ozone, a greenhouse gas. Despite these interrelationships among pollutants, examination of issues and the making of regulations has, for the most part, occurred without serious consideration of the linkages. In most instances, ozone and oxidants, SO_2 and sulfates, acidic deposition, and fine particles have been examined separately. The exceptions include linking acidic deposition and sulfur oxides since 1980 (SO_x was treated alone before that time) and linking visibility impairment and fine particles.

Proper attention to the interrelationships can influence the development of long-term emissions control strategies. For example, a reluctance to reduce NO_x emissions in order to attain the ozone standard might be countered by a desire to reduce NO_x to lessen acidic deposition or formation of fine particles. One might speculate that, were interpollutant issues treated as such through the 1980s, attention would have been given to the merits of NO_x control at an earlier date.

In the late 1960s, the NAPCA identified ozone and oxidants, CO, particulate matter, hydrocarbons, SO_2, and NO_2 as criteria pollutants. "Criteria documents" were drafted for each. State or federal implementation plans generally addressed control strategies separately for each secondary nonattainment pollutant (i.e., a pollutant not directly emitted, but formed in the atmosphere through chemical reaction) and its precursors. In effect, this artificially disentangled one reactive system from the other. Existing links generally received only passing attention, or they were ignored. This approach persists today.

No rules prevent an agency from preparing an SIP or FIP that examines and respects the linkages. Even though a historical precedent has not been established, the vehicle for such examination exists. Both *Catching Our Breath* (OTA, 1989) and Title IX of the 1990 amendments recognize the need to establish links and interactions. The OTA notes that "NO_x emissions affect more than just nonattainment area ozone concentrations, further complicating the decision about whether to mandate controls. NO_x emissions contribute to acid deposition." Title IX directs that the administrator of the EPA "shall conduct a program of research, testing, and development of methods for sampling, measurement, monitoring, analysis, and modeling of air pollutants. Such program shall include . . . consideration of individual, as well as complex mixtures of air pollutants and their chemical transformations in the atmosphere . . . and interactions of ozone with other pollutants." Here, recognition of the need and the appropriateness is expressed in a research prescription. The findings of the research will obviously be applied only over the longer term.

The potential for formally examining interrelationships has existed for some time. Knowledge gained from the study of individual groupings of pollutants is certainly applicable. Recognition of the desirability of examining the complex interrelationships among pollutants is reflected in the formation of the Consortium for Advanced Modeling of Regional Air Quality (CAMRAQ) (Hansen et al., 1992). This group consists of public agencies and associations representing research interests of the private sector. The CAMRAQ was formed in 1991 at the suggestion of the Electric Power Research Institute (EPRI). It subscribes to the following principles:

• Air pollution occurs over a range of spatial scales, and its study should reflect this attribute.

• Air pollutants are linked chemically in the atmosphere, and the study and representation of their reactions should be appropriately comprehensive.

• The problems are the concern of virtually all public agencies and private sector interests with operations that emit substantial amounts of air pollutants.

• A consortium may prove to be a convenient means for establishing research needs and funding the requisite efforts.

• Large-scale modeling using high-speed advanced computing techniques will be essential to this pursuit.

The EPA is an active member of CAMRAQ; it has also recently initiated its own program of advanced, interlinked modeling.

Because both the CAMRAQ and EPA programs are in their formative phase, it is difficult to anticipate the full nature of the programs to be carried out, the relevant levels of effort, or the schedules for developing products and findings.

Spatial Scales

Pollutant processes at the regional and urban scales can be highly interactive: emissions and air quality in surrounding regions can contribute significantly to ozone formation in urban areas, and ozone formed in urban areas can contribute significantly to regional ambient ozone concentrations. Accounting for the transport of ozone and its precursors to downwind nonattainment areas is a problem that plagues the development of strategies for attainment. Field studies carried out in the late 1970s and the early 1980s identified transport as a significant issue for a number of nonattainment areas; however, only recently has it been addressed in amendments to the Clean Air Act. In this section, we briefly review the understanding of ozone transport and discuss the extent to which legislation, regulation, and guidance have addressed the issue. Two "transport-related" questions are central to developing effective regional ozone strategies:

• At what spatial scales is effective planning for pursuing attainment of the ozone standard best achieved?
• For larger and more heterogeneous regions, should spatial variability in emissions control requirements be introduced? If so, under what circumstances?

Before the early 1970s, little, if any, attention was given to establishing the potential role of transport in ozone formation. Since 1974, measurement programs carried out in the Midwest, the East, and California have identified urban plumes and regional (ozone) background of anthropogenic origin as contributing significant amounts of ozone and precursors (mainly aged hydrocarbons) to downwind urban nonattainment areas under "favorable" meteorological conditions. In the aggregate, these studies characterized conditions conducive to transport and monitored concentrations of transported pollutants at ground level and aloft.

The Midwest Interstate Sulfur Transformation and Transport Study (MISTT), conducted in the mid-1970s, was among the first studies to identify transport as an issue (White et al., 1976). Investigators determined that ozone was formed in the pollutant plume originating in St. Louis. They were able to map its size and shape and estimate its distance of transport. In 1979 the Northeast Regional Oxidant Study I (NEROS I)—a field study in which aircraft were used to carry out measurements—demonstrated the transport of ozone from Ohio to the East Coast (Clark and Ching, 1983). In addition although they were primarily aimed at acidic deposition, field measurement and air quality modeling efforts that were a part of the National Acid Precipitation Assessment Program (NAPAP) identified ozone transport paths and the meteorological conditions associated with pollutant transport in the eastern United States (NAPAP, 1991).

In California, studies designed to document transport included tracer and aircraft measurements between the Sacramento River Delta and the San Joaquin Valley (1976), within the San Joaquin Valley (1979, 1984, 1990), and between the San Francisco Bay Area and the valley (1990). The 1974 Aerosol Characterization Experiment (ACHEX) and 1981 Southeast Desert Air Basin Study (SEDAB) provided evidence of transport from the Los Angeles Basin to the southeast desert area. Two studies documented transport between Los Angeles and Ventura (1983) and Los Angeles and Santa Barbara (1985), and identified associated meteorological conditions. By the mid-1980s, regional transport was broadly recognized as a principal contributor to nonattainment of the ozone standard in many parts of the country.

As noted, the Clean Air Act amendments adopted through 1970 did not specifically address transport nor did they give much attention to nonattainment planning. With the passage of the 1977 amendments, two changes occurred. First, the act recognized transport. However, emphasis was given to emissions from individual sources located in attainment areas and their impacts on air quality in nonattainment areas located across state boundaries [Sections 110(a)(2)(E) and 126]. Second, the act specifically defined the requirements for the state implementation plan for ozone by adding Part D to Title I, which addresses stationary source permit and planning issues. Still, neither the act nor subsequent regulations required that planning specifically account for transported pollutants. In fact, guidance through the 1980s generally led states and local districts to select ozone episodes involving minimum transport for modeling and planning purposes. Thus, agencies tended to restrict planning to examining local impacts of local emissions (personal communication with A. Ranzieri, 1992; personal communication with R. Scheffe, 1992).

An approaching December 1987 deadline for ozone attainment led the EPA, concerned with future SIP efforts, to address issues associated with transport and more extended spatial scales. The agency released a document entitled, "Post-1987 Planning Guidance" (FR, Vol. 52, November 25, 1987), which extended earlier guidance to include knowledge gained from the review of the latest planning efforts. The document recommended that planning studies using photochemical models take into account sources that were up to 25 miles upwind of a CMSA. Emissions from sources farther upwind were to be reflected in pollutant concentrations at the upwind boundary. However, no specific requirements for control of emissions from these upwind sources were developed apart from those needed to attain and maintain the ozone standard locally because no regulations had ever been issued to support Sections 110 (a)(2)(E) and 126.

Two significant technical impediments forestalled development of the information needed to formulate regulations. First, the EKMA (the model recommended by EPA for use in SIP development) had a number of shortcomings that seriously limited its applicability to assessing control strategy options. Although the model contained a sophisticated chemical mechanism, its treatment of meteorology was overly simplified. Transport from upwind areas could not be adequately characterized. Moreover, as a result of the assumptions on which it was based, the model was not appropriate for simulating multiday episodes, which precluded using it to assess transport at distances greater than that covered in a travel time of about 10 hours between the initial location of an upwind air parcel and downwind areas of interest. Second, there was, and continues to be, a paucity of data on air quality, meteorology, and emissions needed to operate and evaluate the EKMA (as well as more sophisticated urban scale photochemical grid models such as the Urban Airshed Model).

The 1990 amendments recognized concerns about transport of ozone and its precursors between urban areas in a region by adding two new sections to the act. Section 176A establishes authority for states or the EPA administrator to form interstate transport commissions empowered to recommend regional approaches to mitigating interstate pollution. As noted earlier, Section 184 specifically addresses control of interstate ozone air pollution and mandates a Northeast transport region consisting of 11 states from Maryland to Maine and the District of Columbia, with the subsequent addition of Virginia. However, none of the mandated requirements in Section 184 addresses the larger issue of regional planning.

Although little technical guidance has been offered in accounting for transport in planning, only limited experience exists upon which to base such guidance. The EPA "Guideline for Regulatory Application of the Urban Airshed Model" (EPA, 1991) supports the necessity of modelling episodes associated with upwind transport as well as episodes that are predominately locally generated. Issues associated with geographical scale are recognized, and the EPA recommends that the spatial extent of an upwind region be set "as large as feasible . . . to reduce the dependence of predictions on uncertain boundary concentrations and to provide flexibility in simulating different meteorological episodes." Even so, the agency recognizes that, as a practical matter, there may not be enough data to extend the upwind boundary sufficiently. Thus, the agency further recommends that states use its regional oxidant model, a coarse-scale, first-generation regional model, to provide upwind boundary concentrations for urban-scale modeling. Nowhere in either this guideline or other EPA documents are there specific recommendations for evaluating the merits of uniform versus spatially variable emissions controls.

The 1988 California Clean Air Act specifically recognized the need to assess the impacts of transport from upwind areas in planning and to minimize its impacts over the longer term. Both the act and subsequent regulations require that upwind nonattainment areas develop control strategies that will attain the state and federal ozone standards and also mitigate (to the extent possible) transport to a nonattainment area located downwind. Control requirements are established for upwind areas; they specify the levels of reductions in VOC and NO_x emissions and the percentage of sources that must be controlled. However, an alternative set of controls that is equally or more effective, as demonstrated by modeling, is allowed. Downwind areas are required to provide the reductions needed to mitigate the exceedance of the standards that would occur if there were no transport (California Clean Air Act, AB 2595, Chapter 1568, Statute of 1988). However, similar to the national situation, specific technical guidance has yet to be developed for carrying out the analyses needed to meet these mandates.

As indicated, the EPA has been slow to develop regulations or guidance for addressing pollutant transport issues associated with ozone. In large part, this has been because there are no specific data suitable for air quality modeling and analysis of long-range transport, activities that would provide the means for determining the appropriateness of alternative action plans. The language of the 1990 amendments does not mandate that the EPA develop a plan encompassing transport issues. Rather, the states comprising an ozone transport commission are empowered to do so, following EPA-issued guidance. These plans are subject to approval by the EPA.

At this time it is quite difficult, unfortunately, to address substantively the two questions posed at the outset. Current knowledge is inadequate. During the 1980s, either the long-term funding needed to support acquisition of data and improvement of models was unavailable, or the proposed studies were judged to be of insufficient priority to receive the necessary financial support. Currently, three comprehensive monitoring and modeling studies focusing on ozone are in progress—one in central California, encompassing the region from the San Francisco Bay Area to the southern end of the San Joaquin Valley, the second in the area encircling southern Lake Michigan, and the third in the southeastern United States. The first two studies were specifically designed to provide data bases to support source-oriented photochemical modeling and will be using a second-generation regional ozone model to evaluate alternative potential control strategies. The findings of these studies should provide valuable information for planning regional ozone attainment strategies.

The basic issues facing policymakers today who are concerned with air quality planning on a regional scale include determining:

• If a mixed or hybrid strategy—for example, VOC control in the urban area and NO_x control in the surrounding rural areas—is likely to be preferred over more traditional uniform strategies,

• The levels of emissions reductions needed in the urban and rural areas,

• The amount of transport to downwind urban centers that is likely to occur under emissions control,

• The impact of that transport on downwind metropolitan areas, and

• The actions that should be taken to mitigate adverse effects in air quality in downwind areas.

Efforts are needed to develop procedures for strategic planning at the regional level. For example, one might elect to evaluate control measures sequentially, starting with the specification of measures for attaining the standard in the nonattainment area farthest upwind. Efforts to determine a preferred strategy for the area immediately downwind would then take into account concentrations of "imported" ozone and precursors. We advocate developing both the needed modeling and strategic planning capabilities as soon as practicable.

Motor Vehicles: Inspection and Maintenance

• Are motor vehicle emissions being effectively reduced through fleet emissions reductions and inspection and maintenance (I/M) procedures?

Four general approaches have historically been taken, or are being considered, to reduce automotive emissions: (1) technological improvements and modifications to motor vehicles to meet tighter tailpipe and evaporative emission standards and requirements for in-use compliance with the standards; (2) vehicle inspection and maintenance programs; (3) use of reformulated gasolines or other fuels that either reduce emissions of pollutants or are less reactive in the atmosphere; and (4) transportation control measures. In this section, we review the legislative and regulatory history of the first two approaches, with emphasis on the passenger vehicle. We highlight some of the issues that have been associated with these approaches.

Transportation control measures are discussed in a later section. We will not examine the third approach because the requirement for reformulated gasoline has only recently been introduced, and it is premature to examine the issue in this paper.

Technological Improvements and Modifications to the Motor Vehicle

Congress, in amending the Clean Air Act in 1970, established emission standards that required a 90 percent reduction of hydrocarbons, CO, and NO_x from the base year of 1970 for light duty vehicles (1971 for NO_x). The standards for CO and VOCs were to be met by 1975 and for NO_x by 1976. Although regulators may have held opinions concerning the means for achieving the specified reductions, the legislation and subsequent regulations avoided prescribing a methodology. Rather, required targets for emissions reductions were set, in effect forcing technological development. However, this development required more time than was allotted in the initial legislation. In August 1972 automakers petitioned the EPA for a one-year waiver, indicating that development of a catalytic converter required more time (Quarles, 1976). The EPA denied the petition, and the automakers filed for judicial review of the decision. An appeals court overturned the EPA's ruling, and additional hearings were held. In April 1973 the EPA granted a one-year delay based on the testimony at the hearings, but at the same time it tightened emissions requirements. Delays were granted two additional times through 1975 for a variety of reasons. Congress, through the 1977 Clean Air Act Amendments, extended the implementation dates in a two-step process to 1981; the standards put in place at that time will remain operative until the 1990 Clean Air Act Amendments are implemented in the mid-1990s. The present standards, given in grams (g) per mile, are 0.41 for hydrocarbons, 3.4 for carbon monoxide, and 1.0 for NO_x, based on federal fleet certification procedures.

Consistent with a philosophy of forcing technology through performance standards, California adopted a program in 1990 that requires the development of transitional low emission vehicles (TLEVs), LEVs, ultra-LEVs and Z(ero)EVs over the next seven years (CARB, 1990). The specified emission limits for nonmethane organic gases (a subset of hydrocarbons) are 0.125 g/mile for TLEVs (1994), 0.075 g/mile for LEVs (1997), 0.04 g/mile for ULEVs (1997), and 0 g/mile for ZEVs (1998). The performance requirements of this program surpass that of 0.25 g/mile specified in the 1990 federal amendments adopted later in the year.

To date, vehicle manufacturers have demonstrated that some of their late models will meet the TLEV standards, and a few TLEVs have been certified. To meet LEV requirements, manufacturers are considering different technologies, such as electrically heated catalysts to reduce cold-start emissions (which are currently the largest source of emissions in the federal test procedure [FTP]), closed coupled catalysts, enhanced catalyst performance, and improved fuel injection. However, the combination of technologies that will be used for the LEVs and ULEVs has not been determined. Furthermore, the actual deterioration rates associated with these new technologies are unknown. Reliable estimates of the extent to which this program will satisfy stipulated targets (emissions limits) and constraints (schedule and costs) cannot be provided now.

Vehicle Inspection and Maintenance Programs

Congress recognized in 1977 that some ozone nonattainment areas might not achieve the standard by the December 1982 deadline, even though all requirements and emission reductions specified in Title I, Part D, dealing with planning and stationary source permits, had been implemented. Part D provided for an extension of the attainment date to December 1987, but, in addition, it required in Section 172(b)(11) that the states "establish a specific schedule for implementation of a vehicle inspection and maintenance program."

Inspection and maintenance (I/M) programs today vary from state to state, but their purpose is to reduce in-use vehicle emissions through identification and repair of vehicles with high levels of emissions. They typically consist of an annual or biennial inspection and measurement of tailpipe emissions at two different engine speeds with no load on the engine. If a vehicle fails the test, it must be repaired and retested before it can be registered. The EPA established specific guidelines for the program and provided estimates of emission reductions that could be taken as part of the SIP reasonable further progress (RFP) requirements. These estimates were calculated using the EPA mobile source emissions factor program, MOBILE (or the California equivalent, EMFAC). States that implemented I/M were allowed to take credit in their SIPs for a 25 percent reduction of tailpipe emissions from mobile sources.

The EPA-mandated I/M program was based on two operating programs that had been conducted by the city of Portland, Oregon, and the state of New Jersey during the mid-to-late 1970s. By the mid-1980s, numerous states had adopted some form of the I/M program and had begun taking credit for the reductions. Unfortunately, as pointed out by McConnell and Harrington (1992) in their recent study of the cost-effectiveness of enhanced I/M, the program is not achieving its goal.

> Simply having a program in place was sufficient for getting the credit for the 25% reduction toward the state's SIP. There were few requirements for enforcement, including no specific checks for tampering. Nor was there any requirement to link I/M credits to enforcement or to actual in-use vehicle emissions (McConnell and Harrington, 1992, p. 5).

Part of the problem lies in the difference between the EPA's design for the program and its practical implementation. The EPA developed the I/M in a laboratory setting. The pass/fail thresholds were chosen on the basis of iterative analyses such that vehicles that passed the laboratory tests should not fail FTP certification. The EPA engineers identified the causes of failure in the test vehicles, made repairs, and retested the vehicles, including operating them over the FTP cycle. Thus, the EPA developed an idealized laboratory I/M program that produced the expected benefits. However, when this program was implemented nationwide, vehicles did not undergo evaluations as intensive as those performed by the EPA. Both this "transference problem" and a failure of the programs to discourage tampering apparently account for the shortfall in emissions reductions.

The present two-pronged automotive emissions reduction program—(1) catalytic conversion of emissions, coupled with precise electronic control of the air-fuel mixture, and (2) post-1982 implementation of I/M—was operative by the mid-1980s. In principle, each part complements the other. The certification standards cover fleet emissions reductions, and the I/M program is intended to capture those vehicles whose emissions performance has degraded through age, neglect in maintenance, or tampering with the control system.

During the past few years, the CARB and the EPA have carried out a number of studies that have analyzed the effectiveness of California's Smog Check Program and examined automobile emissions under actual operating conditions. Among the states, California has taken the lead in assessment; we thus focus on conclusions drawn from CARB-sponsored studies.

Initial and subsequent legislation for the California I/M program required evaluations of the effectiveness of the program. These evaluations were carried out by CARB through covert vehicle emissions testing programs in which vehicles identified as high emitters were taken to I/M testing stations to determine if they passed or failed. The vehicles were then inspected by the CARB's automotive testing lab to ascertain if repairs made by the stations were appropriate. In 1986, 800 vehicles were evaluated, in 1989, 1,100 vehicles. The results of the more recent evaluation program have established that, taking into account deterioration effects and the residual benefits of previous inspection cycles, the Smog Check Program has reduced hydrocarbon emissions 19.6 percent and carbon dioxide emissions 15.3 percent (California I/M Review Committee, 1992). These reductions are less than anticipated for the I/M program and less than those required by state law.

While the I/M evaluation studies provided information on the actual effectiveness of the program, other studies analyzed measurements made under actual operating conditions. In one study (Pierson et al., 1990), follow-up analyses compared measurements made in the previously cited Van Nuys tunnel study with vehicle emissions data. It found that the measurements

> gave higher CO and HC emission-rates values than expected on the basis of [estimates of an] automotive-emissions model by factors of approximately 3 and 4, respectively (Pierson et al., 1990, p. 1485).

Pierson et al. offered the following observations in their recommendations:

> Accordingly, it becomes important to verify that this [large discrepancy] is so and to understand why-whether air/fuel ratios are richer in on-road operations than has been thought, whether evaporative emissions have been incompletely accounted for, whether tampering and/or maintenance is worse than assumed, and so forth (p. 1495).

On the basis of the analyses of the Smog Check Program, the CARB has concluded that the present I/M program estimates of actual tailpipe reductions are too high, and that evaporative emissions are not being taken into account. Calculations made using the revised CARB mobile source emissions estimation model, EMFAC-7EP, suggest that emissions reductions of 18 percent for CO and hydrocarbons will provide appropriate credits for the present I/M program.

The 1990 Clean Air Act Amendments required that states enact an enhanced I/M program for areas which are classified as serious or worse under the ozone nonattainment designation requirements. The program is to also include

a performance standard achievable by a program combining emission testing, including on-road emission testing, with inspection to detect tampering with emission control devices and misfueling.

In July 1992, the EPA proposed an enhanced I/M program (*FR*, Vol. 57, July 13, 1992) that will require a dynamometer test, with the vehicle completing the first 240 seconds of the hot start FTP cycle under varying engine loads. The program will also be capable of testing for NO_x as well as hydrocarbons and CO. The proposed rule was finalized in November 1992. One very important part of this new program, which is missing from the present I/M program, is the ability to detect failure of the evaporative emissions control system. Evaporative emissions have been identified by both EPA and CARB as a significant portion of the overall mobile source emissions inventory, but these emissions were properly taken into account only in the most recent versions of EPA's MOBILE emissions model. The EPA estimates that the enhanced I/M program will provide significantly greater emission reductions (28 percent for hydrocarbons, 30 percent for CO, and 9 percent for NO_x) than the present program.

California has taken the lead among the states in examining vehicle emissions. As noted above, California legislation established the I/M Review Committee. Part of the legislation required this committee to include recommendations in its 1992 report to the legislature on "the most effective means of reducing tampering and emissions control equipment failures which result in high-emitting vehicles."

This legislative mandate was motivated by several recent CARB studies that concluded that approximately 10 to 20 percent of vehicles produce between 40 and 50 percent of total vehicular emissions. The California I/M Review Committee report has identified several different options that may be available in the near future to identify these high-emitting vehicles so that they may be repaired. Each of these options requires some technological development and costs and effectiveness will need to be analyzed before any one is considered for implementation. The potential options are: (1) use of remote sensing to detect high tailpipe emissions, followed by testing of the vehicle, (2) random roadside surveys in which a small percentage of vehicles would be tested, and (3) on-board diagnostic systems that could detect tampering through a radio transponder implanted in the system.

In summary, both the 1977 mandated emissions control technology and the I/M program have provided for significant reductions in vehicular emissions. The 1990 Clean Air Act Amendments will continue to mandate improvements in each program as a means to further decrease automotive emissions.

Motor Vehicles: Reducing Use

• To what extent should measures that directly or indirectly limit vehicle use be considered in formulating regulations for reducing ambient ozone concentrations?

During the past two decades, the public's consciousness of the adverse impacts of urban growth has increased in proportion to the diminution in quality of life, on a day-to-day basis, that the public confronts. Road construction has promoted suburban sprawl. Increased suburban populations have resulted in longer commutes, increased roadway congestion, and reduced average speeds on highways. Air quality is adversely affected as well. While these issues go far beyond air quality concerns, some measures which address these issues, such as transportation control measures (TCMs), have been justified on air quality considerations alone.

As we have pointed out, present federal environmental laws tend to focus on specific issues, e.g., reducing the impacts of a particular pollutant by reducing emissions from a category of sources. The framework within which congressional decisions are made is often similarly constrained. More integrated or comprehensive approaches to planning and regulation have occurred only infrequently. An increasing awareness of the limitations of depending on an "issues-oriented" strategy alone to meet objectives has led to the advancement of more encompassing recommendations. These include proposals for restricting land use at the local and regional levels—in effect, promoting limits on growth and development.

Proposals focusing on mitigating the adverse effects of air quality associated with land use include measures for reducing automobile use, either directly through TCMs or indirectly through the review and approval of new construction projects, taking into account anticipated increases in vehicle mileage traveled (VMT) that a given project is expected to induce. Examples of TCMs include increased bridge, tunnel, and highway tolls; mandated increases in vehicle occupancy; "no drive" days; restrictions on driving and parking in center city; and air quality permits for construction of parking lots. Types of "new construction" projects include industrial facilities, shopping centers, and residential developments. The permit process that results in approval or disapproval of a construction project and defines mitigation measures to decrease the anticipated VMT associated with the project is termed indirect source review (ISR).

As discussed, in 1971, the EPA published regulations governing preparation of the first state implementation plans for oxidants (*FR*, Vol. 36, November 25, 1971). Reducing automobile tailpipe emissions and controlling new stationary sources were the principal means for attaining the standard. However, the EPA encouraged states to develop TCMs and consider permits for indirect sources. Among the TCMs that the EPA suggested were the following:

> measures to reduce vehicle traffic, including but not limited to, measures such as commuter taxes, gasoline rationing, parking restrictions, staggered work hours . . . [and] expansion or promotion of the use of mass transit facilities through measures such as increases in the frequency, convenience, and passenger carrying capacity of mass transportation systems or providing for special bus lanes on major streets and highways (*FR* 36, 22398, November 25, 1971).

The EPA did not specify the approaches to be taken in controlling emissions resulting from indirect sources, but it did not rule out land use measures.

Later in the year, the agency enacted a national program for indirect source review and a Transportation Control Plan (*FR*, 36, November 25, 1971). The Transportation Control Plan contained the TCMs suggested in the 1971 SIP regulation. The ISR program had the intent of approving or denying a permit for the proposed indirect source on the basis of its air quality impacts. This program gave rise to numerous lawsuits throughout the United States. Although individual legal challenges had different results, the overall outcome established that the EPA lacked the legal authority to compel the states to adopt these measures (personal communication with M.R. Barr, 1992). The EPA withdrew the ISR program in 1975 as a result of congressional action (*FR*, 40, 129, July 3, 1975).

In 1974 Congress made minor amendments to the Clean Air Act in the Energy Supply and Environmental Coordination Act. This act specifically prohibited the EPA from promulgating or requiring the states to promulgate an indirect source review program and left the program to the discretion of the states. Further, Congress redefined TCMs in light of the earlier problems associated with the EPA's Transportation Control Plan by removing the authority to require implementation and leaving adoption and implementation to the discretion of the state. Examples of TCMs specifically cited in Section 108(f)(1)(A) of the 1977 amendments are the following:

> programs to improve public transit, programs to establish exclusive bus and car pool lanes and area wide car pool programs, programs for long range transit improvements involving new transportation policies and transportation facilities or major changes in existing facilities.

As a result, although all pre-1992 SIPs retained TCMs, they tended not to take credit for the estimated emissions reductions that were to derive from implementing the measures, in part because of the voluntary nature of the program and in part because of the difficulty in estimating the likely air quality improvements. Even though attempts to decrease VMT through the use of TCMs were unsuccessful, the question remains as to whether emissions reductions related to TCM and land use are significant.

In recent years, Congress has attempted to enact more stringent legislation involving TCMs. The TCM-related provisions of the 1990 amendments aim to reduce emissions by requiring that ozone nonattainment areas classified as severe and extreme develop plans to offset growth in emissions due to VMT. As interpreted by the EPA, changes in the fleet composition and other measures that reduce tailpipe emissions can be credited toward maintaining total vehicle emissions at a constant level. The plans, which were to be submitted as part of the November 1992 SIP revisions, were to emphasize transportation control measures that are identified in Section 108(f)(1) of the amended act. These measures include such programs as improved public transit, high occupancy vehicle lanes, trip reduction ordinances, restriction of vehicles in downtown areas during peak hours, and employer-based transportation management plans. Many of these programs require enabling legislation at local, regional, and state levels. In addition, the 1992 SIP revisions must contain compliance plans for each employer of 100 or more persons to increase average passenger occupancy of vehicles in which employees commute between home and work by 125 percent by November 1996.

Few states have tried to quantify emissions reductions that will result from implementation of the 1990 amendments. However, the 1988 California Clean Air Act contains requirements for reducing VMT similar to those of the 1990 amendments. As part of the 1991 Clean Air Plan, the Bay Area Air Quality Management District (BAAQMD) has estimated the benefits of these reductions. They are small, amounting to 10 tons/day (t/d) of hydrocarbons (2 percent of the projected baseline 1997 emissions inventory) and 16 t/d of NO_x (2.8 percent of the projected inventory). The district has also adopted additional TCMs, including some measures proposed by the EPA in 1971. These include smog fees, congestion pricing, work parking charges (i.e., institution of the practice in which employers must charge employees for parking privileges at their work place), and gas tax increases; most of these will require additional legislative authority before they can be implemented. Even these measures, combined with the previous reductions, are estimated to result in only a 41 t/d decrease in hydrocarbons and a 57 t/d decrease in NO_x. The BAAQMD has ranked all of the measures adopted in the 1991 Clean Air Plan in terms of cost-effectiveness. Except for a single measure to expand employer assistance programs (ride sharing), all transportation control measures have gross costs of between $25,000 and $1 million per ton of emissions reduced. These costs are much higher than the majority of those for stationary and area source measures cited in the plan (BAAQMD, 1991).

The BAAQMD cost estimates raise important questions as to whether TCMs are cost-effective measures for improving air quality. Independent estimates from other districts and states are needed before one can draw general conclusions about the cost-effectiveness of TCMs. That information should be forthcoming at the time of the 1994 SIP submissions. Nevertheless, state and local governments may elect to adopt indirect source measures and TCMs, including those TCMs considered in the past (which, in some cases, have proven to be controversial). Where such actions are considered, it will certainly be appropriate to assess their air quality benefits through quantitative analysis.

In summary, policies that support indirect source review and TCMs are more likely to be justified through their contribution to alleviating the local and regional societal concerns that were noted in the beginning of this section rather than through their air quality benefits alone.

Other Topics

This section summarizes topics of interest that could not be included in the more detailed preceding review.

Assessment of Progress Toward Attainment of the Standard

The ozone monitoring network in the United States provides an adequate, although not fully satisfactory, sampling of ozone concentrations in and immediately downwind of urban areas. Rural monitoring is more sparse, and the network would certainly benefit from bolstering. However, VOC and NO_x, the precursors to ozone, are not measured, or are measured sparingly, in most urban areas. This paucity of information severely limits the ability to evaluate the effectiveness of emissions control programs.

Two reasons appear to account for this deficiency in monitoring. First, the technology available for measuring VOCs and NO_x was, until recently, not accurate enough. Moreover, acquisition and accurate analysis of VOC samples are costly; typically, costs are $300 to $500 per sample. (The development of inexpensive, continuous monitors for VOCs is still needed.) Second, the regulatory community appears to have done one of two things: (1) It has accepted the notion that the costs saved by not monitoring precursors outweigh the benefits of the additional information, which would be useful in establishing trends, assessing the effectiveness of control programs, and verifying the accuracy of emissions models and inventories. (2) Or, it has overlooked, or been unaware of, the importance of having such

information. In this case, added information would have promoted the use
of science to support the regulatory process. Unfortunately, those scientists
who were sensitive to the need for such information for many years were
clearly unable to convince federal and state agencies of the importance of
conducting the needed monitoring.

Retrospective Analyses of State Implementation Plans

Historically, few, if any, SIPs have provided accurate estimates of improvements in air quality related to ozone levels. Virtually all estimates of
future air quality have proven to be overly optimistic; actual peak ozone
concentrations in retrospect have been higher than those projected. The
API, in a 1988 analysis of ozone SIPs, found that 1982 SIPs suffered several generic deficiencies that caused this problem, including underestimation of base year emissions inventories and the resultant underestimation of
future year inventories, failure to apply the EKMA model only in conditions
in which its assumptions are met, overestimation of effectiveness factors for
emissions reduction measures, and inadequate representation of long range
transport of pollutants into a region (API, 1989). The overestimates of
effectiveness were the result of delays in developing rules, inaccurate estimation of actual "technical" effectiveness, and lack of compensation for
shortfalls in emissions reductions (API, 1989).

While SIPs have historically displayed common patterns of inaccuracy,
methods of detection and correction have been inadequate or lacking. Critical elements include monitoring designed to detect and quantify progress
toward attainment, comparison of progress with SIP projections, identification of shortfalls in air quality improvement, determination of the causes of
shortfalls, specification of corrective steps in control, revision of the SIPs to
reflect the needed changes, and implementation of the controls. "Science"
has recognized this need; regulation has not.

Introduction of "Cleaner" Fuels

Title II of the CAAAs of 1990 requires the eventual sale of only reformulated gasoline in the nine urban areas having the highest peak ozone
concentrations, institutes two clean-fueled vehicle programs, including a
pilot program in California, and mandates maximum fuel volatility. While
volatility reductions have led to reduced VOC emissions, the impact of
introducing "cleaner" fuels is less clear. Today, many argue that the legislation requiring the introduction of clean fuels preceded proper development
of the supporting science. Some believe that the long-term benefits are
questionable. Whatever the eventual findings, "clean fuels" is a major issue
today, one that certainly merits attention and scrutiny. We did not address

the topic in this paper because of its relative novelty and thus lack of history, the extreme sensitivity of the issue, and the extensive coverage that the topic would necessarily have to receive. However, the clean fuel and reactivity issues are of at least the same interest and timeliness as most of the topics addressed earlier.

THE INTERPLAY OF REGULATION AND
SCIENCE AND TECHNOLOGY DEVELOPMENT

Scientific and technological advances and legislative and regulatory history are intimately intertwined in our earlier discussion. We now examine the interplay among them and develop some general observations about the influence of each on the other. Figure 3 provides a timeline of key events.

The Response of the Regulatory System
to New Scientific Understanding

A general observation emerging from this inquiry is that the regulatory system has often responded to new scientific understanding of tropospheric ozone either with inertia or only after considerable delay. At times, there have been good reasons for the lag time. Steps in the process can be characterized as follows:

• Advances in knowledge develop gradually over time.

• As knowledge accumulates, existing views or positions become subject to challenge, the vigor of which increases in proportion to the growing evidence.

• Uncertainties persist, making it difficult to characterize adequately weaknesses or flaws in current and, sometimes, prior positions.

• Interests on each side of an emerging or persistent issue tend to represent their positions in committees or public forums by emphasizing the pros and minimizing the cons, which often results in polarization.

• Because of the inability to establish an incontrovertible position, debate continues for some time, leading to delay.

• Eventually, the new position or view holds sway, and change is introduced.

Although the time period associated with this process is variable, it seems to range from 5 to 15 years.

Earlier we discussed issues that exemplify this pattern:

• *Revisiting specification of the ambient standard for ozone.* The Clean Air Act provides for review of the ozone standard at five-year intervals.

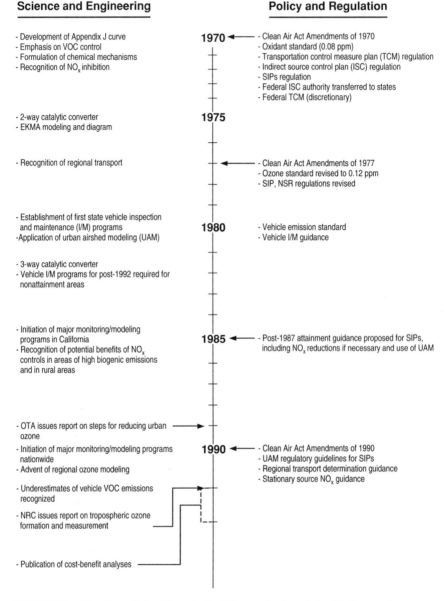

FIGURE 3 Timeline of significant scientific and technical, legislative, and regulatory events for ozone control.

The EPA announced in August 1992 that it will not alter the ozone standard at this time, stating that it did not have enough time to compile information describing relevant study efforts and complete its formal review. This announcement came seven years after the 1985 statutory deadline for review.

Although the CASAC had been divided on the need for imposing a more stringent standard when it last met in 1988, Morton Lippmann, who chairs the EPA Advisory Committee on Indoor Air Quality and Total Human Exposure, stated that "new published data since 1988 was available and could have been reviewed in time. They (EPA) recognized if they incorporated the current data, they would no longer be capable of defending the current standard." Bernard Goldstein, director of the Environmental and Health Services Institute at Rutgers University, added that "the additional data that has come out would certainly lean toward making things more stringent" (Weisskopf, 1992). Apparently, both Dr. Goldstein and Dr. Lippmann believe that health-related evidence supports a more stringent standard. In response, Robert Brenner, head of the EPA's Air Policy Office, said that "our focus has been getting controls in place. Putting in new standards doesn't mean you improve the air. It's the regulations we're issuing under the Clean Air Act that will improve air quality."

• *Emphasizing reduction in emissions of both precursors to ozone.* The EPA's long-standing policy favoring VOC control and deemphasizing or disregarding NO_x control, though challenged at various times, remained essentially unaltered from the early 1970s to the late 1980s. Uncertainty clearly prevailed during this period, but even as evidence mounted suggesting the benefits of NO_x control in at least some areas of the United States, the agency was essentially unresponsive until very recently.

• *Recognizing the relationships of ozone precursors to pollutants other than ozone.* Even though a number of studies—from large-scale field programs to detailed modeling efforts—have demonstrated the interactions among air pollutants, the perceived complexity of undertaking comprehensive analyses contributed to inadequate attention being given to the issue. Also, issues can usually be addressed and managed more easily when they are compartmentalized than when they are integrated. The EPA's apparent avoidance of forging integrated pollutant programs may thus derive from practical, albeit untested, concerns.

• *Specifying the spatial scale for planning emissions reductions.* Long-range transport of precursors and secondary pollutants has been recognized clearly for many years. The primary reason for delay in undertaking assessments and planning at the appropriate spatial scales appears to have been political; planning is carried out by designated agencies whose authority is constrained in geographical extent. A significant will is required to effect change; jurisdictions are often unwilling to relinquish authority even if the cause is well intended.

• *Conducting ambient monitoring of precursors to assess and diagnose problems.* Monitoring of ozone concentrations has been supported by regulatory agencies for many years. However, much less attention has been given to the routine measurement of NO_x and VOC concentrations. The paucity of these data has limited the scientific and regulatory communities' ability to determine the effectiveness of emissions reduction programs, the accuracy of emissions representations, and the geographical extent of NO_x limitation.

• *Conducting large-scale integrated modeling and monitoring programs.* While it was well known that it was necessary to acquire comprehensive aerometric data bases to properly support urban-scale and regional modeling, serious commitments of funds to this pursuit were made only in the mid-1980s and subsequently. Before that time, only limited data bases, often compiled using routine monitoring data, were available for use as inputs to air quality models and in evaluating model performance. Thus, the accuracy and precision of modeling estimates were limited by the paucity of data and not by the formulation of the model. This limitation persists today, as the programs launched five to eight years ago are only now reaching fruition; evaluations of model performance using the data bases acquired are just now being conducted. Even today, virtually no regional-scale data bases exist.

• *Characterizing and quantifying the role of natural emissions in ozone formation.* As was discussed earlier, for much of the 1980s EPA's attention was diverted from the study of biogenic emissions in the belief that they were not a significant contributor to ozone formation. Although this view was disputed by some members of the scientific community, progress in this field was effectively limited by the paucity of funding for research to develop methods for sampling and analyzing naturally emitted and highly reactive VOCs, determine the rates of emissions from biota, and estimate total biomass for different species.

In some cases, the delay in acting on new information has been attributable to Congress; 13 years elapsed between 1977 and 1990, the year in which the most recent amendments were enacted. Congressional discussion and debate seemed interminable to many. The need to resolve failures to meet attainment deadlines and to empower the EPA to reestablish the SIP process apparently provided the inducement for resolving the deadlock. To be sure, the implications of acting on several of the key issues were quite significant—costs of the programs; impacts on various cohorts of the population, including jobs; uncertain consequences of introducing new programs, such as emissions trading; and the inevitable dilemma of resolving differences among competing interests. Nevertheless, the process of enacting legislation seems unnecessarily inefficient and prone to favor political over scientific considerations.

In other cases, the EPA has appeared slow to respond to the development of new information. As discussed, the VOC versus NO_x debate has been going on for well over a decade. However, evidence began to build in 1985 that some regions of the country were very likely "NO_x-limited." Guidance emerging from the agency in the late 1980s represented a gradual shift in policy, suggesting that NO_x control might be warranted in some circumstances. However, in post-1987 SIP analyses, the need for NO_x control had to be demonstrated prior to its acceptance as part of a control strategy. In effect, NO_x was "innocent until proven guilty." This position clearly changed with the 1990 amendments; NO_x control is required in areas classified as serious, severe, or extreme unless it can be demonstrated that NO_x reductions do not result in ambient ozone benefits. Now, NO_x is presumed "guilty until proven innocent." This transition in position took about five years to effect. More recently, in December 1991, the NRC Committee on Tropospheric Ozone Formation and Measurement stated that "NO_x control is necessary for effective reduction of ozone in many areas of the United States" (NRC, 1991, p. 11). As a result, the EPA and other agencies are giving intensive attention to evaluating the merits of NO_x emissions reductions.

Similar "case histories" can be outlined for the need to (1) assess the influences of ozone precursors on other regulated pollutants, rather than consider the ozone-NO_x-VOC system in isolation, and (2) examine emissions control requirements on a sufficiently broad spatial scale to include all significant source and receptor areas. In both instances, recognition developed in the late 1970s to early 1980s and evidence of the need increased in the years following. Still, in the former case, no guidance for comprehensive "interactive" assessment is forthcoming. In the latter case, regional studies are now being undertaken in some areas—southern Lake Michigan, central California, and the Southeast. However, a number of geographical areas for which implementation plans are to be prepared are still circumscribed by geopolitical boundaries that do not encompass critical source areas. Consequently, the impacts of and responses to longer range transport of pollutants cannot be effectively addressed in these areas. Note that the EPA has yet to provide guidance on methods for developing control strategies in regions where transport is an issue. Such guidance should address the matter of uniform versus nonuniform strategies, that is, applying the same control strategy throughout the region or varying strategies from one major source area to another within the region.

Finally, some of the apparent delay in response derives from a discomfort with acting, or an unwillingness to act, in light of uncertainty. "Excessive uncertainty," in turn, may be a consequence of governmental research programs not being sufficiently long term and consistently focused to provide the information needed to reduce uncertainties to acceptable levels. If true, this supposition suggests the need for a critical assessment of procedures for identifying research needs, establishing priorities, committing to the long term, and providing sufficient funding to ensure success.

Incentives and Disincentives in the Regulatory System

The primary incentive to seek new information overlaps with the primary disincentive. On the one hand, true knowledge provides the only meaningful basis for actions that circumstances appear to demand. On the other hand, most relevant research requires money and time, often in the range of 5 to 10 years or more. Where circumstances require action, such as smog conditions in the South Coast Air Basin, the waiting time for research results exceeds the time practically available for taking the action(s). Resolution of this dilemma is exceedingly difficult. One option is to design actions that can be carried out in sequence, instituting more stringent controls with time, as needs warrant. The results of research can then influence the "action sequence" as they become available. In any event, inadequate or sporadic funding, disrupting or weakening an otherwise attractive research program, can prove to be a significant disincentive to pursue the activity and thus to seek new information.

Acceptance of New Information by the Regulatory Community

Factors contributing to acceptance or lack of acceptance of new information are complex—part institutional, part psychological, part related to risk averseness, part to commitment to current paradigms. The notion of "acceptance or lack of acceptance" is an oversimplification; often, acceptance constitutes a slow process of learning and becoming comfortable—in effect, a protracted transition. We can only speculate on the factors that are most influential in determining an individual's inclination to use new information. Factors that contribute to resisting the acceptance of new information include those listed below:

• Doubt as to its reliability or correctness.
• Extent to which it conflicts with one's current view of the matter (or the difficulty involved in permitting or accepting the overturning of a current conception or framework for action).
• The perception of intent to delay the taking of an action.
• Averseness to risk, that is, it is often easier to maintain "status quo" than to accept or effect change.
• The need to alter or expand jurisdictional responsibility for regulation (such as expanding a jurisdiction from intrastate to interstate).
• Inadequacy in communication.
• Inability to characterize uncertainty.

In situations in which more than one of these factors comes in to play, such as (1) revisiting the ambient standard, (2) determining if NO_x control is an appropriate strategy, (3) addressing interpollutant issues, and (4) addressing transport considerations, acceptance of a "new position" may be difficult to effect. One might argue that information sufficient to resolve key technical uncertainties seldom becomes available within the time frame in which policymakers feel compelled to act. While knowledge may be incomplete, the latest available scientific information provides the basis for moving forward, weighing options, and promoting change. However, "an adequate knowledge base"—that required to support truly informed decision making—is the pot at the end of the rainbow: while we may try, we never seem to get there.

The job of the decision maker is to evaluate contrasting risks: (1) taking action based on "insufficient data" and risking society's incurrence of unnecessary costs if expectations of benefit are misplaced or unduly optimistic, and (2) delaying action and thus exposing society to greater ambient concentrations of air pollutants than if controls were imposed. One path to easing the policymakers' burden is to develop estimates or at least clear qualitative statements of the risk associated with each option for action or inaction, and communicate this information clearly to decision makers for their consideration. We advocate that suitable procedures for assessment and communication of risk be developed (where they are not now available), prescribed, and applied as soon as practicable.

Governmental Perspectives, Incentives, and Disincentives

Regulatory structures at different levels of government may respond differently to a perceived issue or need. It is not possible to characterize the differences accurately or comprehensively because various governmental structures exist in the United States. However, some general observations can be made.

• Lower levels of government may find it easier to rely on higher levels of government to develop regulations. Often, issues are controversial. Where the "heat is high," it is expedient to point to a requirement placed on a community by "a higher authority" than to enact rulemaking within the community itself.

• Most states follow the federal lead. Some act reasonably promptly; others appear to move slowly and reluctantly. In a few instances, a state will assume a lead role or perhaps chart a somewhat independent course. California is notable in this regard.

The relationship of local or regional development of regulations to state development of regulations bears a resemblance to that of state regulation and federal regulation. Local agencies generally act in response to directives issued at a higher level—sometimes promptly, sometimes slowly. In some instances, a local or regional agency will assume a lead role, particularly in planning or rulemaking. Notable here and familiar to us are the actions of the South Coast and Bay Area Air Quality Management Districts.

• Local perception that a problem is severe and thus in need of attention and a significant level of local support for taking action appear to be requisites for a local or regional agency to assume a leadership role in rulemaking or regulation.

• Where California has assumed a leadership role, it seems to have encountered fewer barriers to action, acceptance, and implementation than has the federal government. One result is that the period from conception to implementation is much shorter. The reasons for the differences between federal and state processes are not entirely clear.

However, Matthew Wald (1992) recently commented on the subject in the *New York Times*. He noted that the California Air Resources Board "has pretty much had its own way and has been what the E.P.A. would probably like to be: tough, well funded and backed by a strong political consensus for cleaner air. . . . It regularly exercises a level of authority that Federal regulators have largely been without since the beginning of the Reagan years." Note that the CARB has final authority in California; EPA positions are subject to review by the Office of Management and Budget (OMB) and, recently, by the White House Council on Competitiveness.

Role of Technology in Shaping Actions of the Regulatory System

As indicated by the history of regulation of tropospheric ozone in the United States, regulations generally prompted the development of the technologies required for compliance. The classic example is that of the catalytic converter for the automobile, which is generally viewed as a success story. Of course, the fundamental science may already have been developed or many or most components of the technologies may already have been available. Regulation focused attention on a particular need and brought together the components required for developing the desired process or product or capability. (Although regulation promoted the development of suitable control technology, it did not specify a technological approach; rather, it prescribed a performance requirement.)

Sometimes science or technology precedes regulation. Grid-based photochemical models were developed during the 1970s; the EPA supported this work to advance understanding and obtain useful simulation capabilities. Later, when the models were judged to be sufficiently advanced, were more widely accepted by the regulatory community, and were viewed as useful, requirements for their application were included in regulation—specifically, in the 1990 amendments to the Clean Air Act. (However, significant issues associated with use of the model, which derive from uncertainties in formulation of the model and deficiencies in supporting data bases, remain to be resolved.) In this case, science led, and regulation followed. In some instances, notably in determining the health effects and threshold concentrations of pollutants, the regulators have sought assistance and advice, have promoted the need for further scientific advances, and have prodded the scientific community to take positions on issues exhibiting significant uncertainties. In such circumstances, the interaction between regulation and science is complex: the roles of scientist and regulator as leader and respondent are not always clear.

A PROPOSAL FOR CONSIDERATION

Several parts of the preceding discussion illustrate difficulties in carrying out longer-term scientific research within a regulatory agency. Science requires time and continuing financial support. Regulatory needs are often shorter term; they frequently demand redirection of effort and reprogramming of funds. Perhaps it would be wise to separate the longer-term pursuit of science from the regulatory structure.

One means for accomplishing this is to create a national environmental research center, funded by Congress. Major research initiatives would be undertaken at the center. Moreover, agencies would be permitted to fund longer-term studies conducted at the center at a level of up to one-third of the center's budget. Short-term reseach would be retained within the agencies. While we realize that many factors must be considered in evaluating the merits of this proposal, we offer the suggestion to promote thought and stimulate discussion.

The interplay between science and regulation merits a much deeper examination than we are able to provide here. The lessons that could be learned, if translated into practice, are likely to justify the effort many times over.

ACKNOWLEDGMENTS

We sincerely thank John H. Seinfeld and Myron F. Uman for their enthusiastic support and encouragement. We also thank John Bachmann, Michael Barr, Don Blumenthal, Ken Demerjian, Basil Dimitriades, Bob Friedman, Fred Lurmann, Will Ollison, Richard Scheffe, Robert Slott, and Steve Welstand for offering their perspectives on historical events, scientific studies, and regulatory actions. We are also most appreciative of Sandra Golding's editing efforts.

REFERENCES

American Petroleum Institute. 1989. Detailed Analysis of Ozone State Implementation Plans in Seven Areas Selected for Retrospective Evaluation of Reasons for State Implementation Plan Failure. API Publ. No. 4502.

Atkinson, R., and A. C. Lloyd. 1984. Evaluation of kinetic and mechanistic data for modeling of photochemical smog. Journal of Physical Chemistry Reference Data 13:315–444.

Bay Area Air Quality Management District. 1991. Bay Area '91 Clean Air Plan. (October 30).

California Air Resources Board. 1990. Resolution 90–58 - Low-Emission Vehicles/Clean Fuels. (September 28).

California I/M Review Committee. 1992. Evaluation of the California Smog Check Program and Recommendations for Program Improvements (Public Draft). (15 October).

Chameides, W. L., R. W. Lindsay, J. Richardson, and C. S. Kiang. 1988. The role of biogenic hydrocarbons in urban photochemical smog: Atlanta as a case study. Science 241:1473–1475.

Clark, J. F., and J. K. S. Ching. 1983. Aircraft observations of regional transport of ozone in the Northeastern United States. Atmospheric Environment 17:1703–1712.

Delucia, A. J., and W. C. Adams. 1977. Effects of O_3 inhalation during exercise on pulmonary function and blood biochemistry. Journal of Applied Physics: Respiratory Environmental Exercise Physiology 43:75–81.

Dimitriades, B. 1972. Effects of hydrocarbon and nitrogen oxides on photochemical smog formation. Environmental Science and Technology 6(3):253.

Dimitriades, B. 1977. An Alternative to the Appendix-J Method for Calculating Oxidant- and NO_2-related Control Requirements. Proc. Internat. Conf. on Photochemical Oxidant Pollution and Its Control, Vol. II. U.S. Environmental Protection Agency. Research Triangle Park, N.C. 871.

Federal Register. 1971. Title 42 - Public Health. Part 410-National Primary and Secondary Ambient Air Quality Standards. 36(April 30):8186.

Federal Register. 1971. Part 51 - Requirements for Preparation, Adoption, and Submission of Implementation Plans. 36(November 25):22398.

Federal Register. 1975. Part 52 - Approval and Promulgation of Implementation Plans. Review of Indirect Sources. 40(July 3) no. 109.

Federal Register. 1979. Part 50 - National Primary and Secondary Ambient Air Quality Standards. 44(February 8):8202.

Federal Register. 1987. State Implementation Plans; Approval of Post-1987 Ozone and Carbon Monoxide Plan Revisions for Areas Not Attaining the National Ambient Air Quality Standards; Notice. 52(November 24):45044.

Federal Register. 1992. Part 51 - Vehicle Inspection and Maintenance Requirements for State Implementation Plans, Proposed Rules. 57(July 13):31058.

Fujita, E. M., B. E. Croes, C. L. Bennett, D. R. Lawson, F. W. Lurmann, and H. H. Main. 1992. Comparison of emission inventory and ambient concentration ratios of CO, NMOG, and NO_x in California's South Coast Air Basin. Journal of the Air and Waste Management Association 42(3):264–276.

Hackney, J. D. 1975. Experimental studies on human health effects of air pollutants. Archives of Environmental Health 30:373–381.

Hall, J. V., A. M. Winer, M. T. Kleinman, F. W. Lurmann, V. Brajer, and S. D. Colome. 1992. Valuing the health benefits of clean air. Science 255:812–816.

Hansen, D. A., R. L. Dennis, A. Ebel, S. Hanna, J. Kaye, and R. Thuillier. 1992. CAMRAQ: The Quest For Modern Solutions to Regional Air Quality Problems. (October 20).

Harley, R. A., A. G. Russell, G. J. McRae, L. A. McNair, D. A. Winner, M. T. Odman, D. Dabdub, G. R. Cass, and J. H. Seinfeld. 1992. Continued Development of a Photochemical Model and Application to the Southern California Air Quality Study (SCAQS) Intensive Monitoring Periods: Phase I. Carnegie Mellon University and California Institute of Technology. Final Report to the Coordinating Research Council under project SCAQS-8.

Ingalls, M. N., L. R. Smith, and R. E. Kirksey. 1989. Measurement of On-road Vehicle Emissions Factors in the California South Coast Air Basin, Vol. 1: Regulated emissions. SCAQS-1, Final Report. No. SwRI-1604. San Antonio, Texas: Southwest Research Institute.

Jeffries, H. E., and R. R. Crouse. 1992. Scientific and Technical Issues Related to the Application of Incremental Reactivity, Part II: Explaining Mechanism Differences. Glendale, Calif.: Western States Petroleum Association.

Krupnick, A. J., and P. R. Portney. 1991. Controlling urban air pollution: A benefit-cost assessment. Science 252:522–528.

Landy, M. K., M. J. Roberts, S. R. Thomas, and V. Nazar. 1990. The Environmental Protection Agency: Asking the Wrong Questions. Chapter 3 in Revising the Ozone Standard (with Valle Nazar). New York: Oxford University Press.

Lawson, D. R., P. J. Groblicki, D. H. Stedman, G. A. Bishop, and P. L. Guenther. 1990. Emissions from in-use motor vehicles in Los Angeles: a pilot study of remote sensing and the inspection and maintenance program. Journal of the Air and Waste Management Association 40:1096.

Linn, W. 1978. Health effects of exposure in asthmatics. American Review of Respiratory Disease 117:835–841.

Lippmann, M. 1989. Health effects of ozone, a critical review. Journal of the Air Pollution Control Association 39(5):676.

Lippmann, M. 1991. Health effects of tropospheric ozone. Environmental Science and Technology 25(November 12):1956.

Melnick, R. S. 1983. Regulation and the courts: The case of the Clean Air Act. Chapter 8 in Air Quality Standards in the Courts. Washington, D.C.: The Brookings Institution.

McConnell, V., and W. Harrington. 1992. Cost-Effectiveness of Enhanced Motor Vehicle Inspection and Maintenance Programs. Discussion Paper QE92–18. Resources for the Future. (April 1992).

National Acid Precipitation Assessment Program. Acid Deposition: State of Science and Technology. 1991. Vol. 1, Emissions, Atmospheric Processes, and Deposition, P. M. Irving ed. Washington, D.C.: Government Printing Office.

National Research Council. 1991. Rethinking the Ozone Problem in Urban and Regional Air Pollution. Washington, D.C.: National Academy Press.

Office of Technology Assessment, Congress of the United States. 1989. Catching Our Breath: Next Steps for Reducing Urban Ozone. OTA-O-412. Washington, D.C.

Pierson, W. R., A. W. Gertler, and R. L. Bradow. 1990. Comparison of the SCAQS tunnel study with other on-road vehicle emission data. Journal of the Air and Waste Management Association 40:1495.

Quarles, J. 1976. Cleaning Up America. Boston: Houghton Mifflin.

Reynolds, S. D., P. M. Roth, and J. H. Seinfeld. 1973. Mathematical modeling of photochemical air pollution - I: Formulation of the model. Atmospheric Environment 7:1033–1061.

Reynolds, S. D., T. W. Tesche, T. Dye, P. Roberts, D. E. Franzon, L. R. Chinkin, and S. B. Reid. 1992. Assessment of Planned NESCAUM/NOTC Modeling Activities, Final Report. Washington, D.C.: American Petroleum Institute.

Schoettlin, C. E., and E. Landau. 1961. Air pollution and asthmatic attacks in the Los Angeles area. Public Health Reports 76:545–548.

Trainer, M., E. T. Williams, D. D. Parrish, M. P. Buhr, E. J. Allwine, H. H. Westberg, F. C.

Fehsenfeld, and S. C. Liu. 1987. Models and observations of the impact of natural hydrocarbons on rural ozone. Nature 329:705–707.

U.S. Environmental Protection Agency. 1971. Air Quality Criteria for Nitrogen Oxides. AP-84, January. Washington, D.C.: Environmental Protection Agency.

U.S. Environmental Protection Agency. 1988. Review of the National Ambient Air Quality Standards for Ozone Assessment of Scientific and Technical Information. Washington, D.C.: EPA, Office of Air Quality Planning and Standards.

U.S. Environmental Protection Agency. 1989. Review of the NAAQS for Ozone: Closure on the OAQPS Staff Paper (1988) and the Criteria Document Supplement (1988). Report of the Clean Air Scientific Advisory Committee (CASAC).

U.S. Environmental Protection Agency. 1991. Guideline for Regulatory Application of the Urban Airshed Model. EPA-450/4-91-013. Washington, D.C.: EPA, Office of Air Quality Planning and Standards.

von Nieding, G. 1977. The acute effects of ozone on the pulmonary function of man. VDI Berichte 270:123–129.

Wald, M. 1992. California's pied piper of clean air. The New York Times 141(3)(September 13):F1.

Wagner, K. K., and N. J. M. Wheeler. 1991. An investigation of modeling emission inventory bias with Urban Airshed Model sensitivity simulations. Presented at Tropospheric Ozone and the Environment II: Effects, Modeling, and Control, Air and Waste Management Association, Atlanta.

Wagner, K. K., N. J. M. Wheeler, and D. L. McNerny. 1992. The Effect of Emission Uncertainty on Urban Airshed Model Sensitivity to Emission Reductions. Presented at Symposium on Tropospheric Ozone: Nonattainment and Design Value Issues, Air and Waste Management Association, Boston, October 28, 1992.

Weisskopf, M. 1992. EPA won't tighten urban ozone standard. Washington Post, August 4, p. A1.

White, W. H., J. A. Anderson, D. L. Blumenthal, R. B. Husar, N. V. Gillani, and J. D. Husar. 1976. Formation and transport of secondary air pollutants: Ozone and aerosols in the St. Louis urban plume. Science 194:187–189.

Keeping Pace with Science and Engineering. 1993.
Pp. 91–140. Washington, DC: National Academy Press.

Municipal Waste Combustion and New Source Performance Standards: Use of Scientific and Technical Information

Suellen W. Pirages and Jason E. Johnston

The release in 1992 of a U.S. Environmental Protection Agency report, *Safeguarding the Future: Credible Science and Credible Decisions* (EPA, 1992), signaled a renewed interest in and concern about application of scientific and technical information in the regulatory process. The development of scientifically based and technically sound regulations requires evaluation of a broad range of factors, including:

• Risks posed if an activity is unregulated;
• Benefits achieved if an activity is regulated;
• The feasibility of controlling risks;
• Costs incurred by a regulated community and the nation both with and without a specific regulatory program; and
• Ranking of risks and costs within national environmental priorities.

Information regarding these factors is continually changing as research efforts within scientific, technical, and medical communities are completed and as national environmental priorities are reevaluated. This case study evaluates the use of technical and scientific information in the development of proposed and final new source performance standards for new municipal waste combustors. Promulgation of these standards was mandated by amendments to the Clean Air Act (42 U.S.C. §7411). Although the focus is limited to performance standards developed for *new* (not existing), large facilities (i.e., greater than 250 tons per day unit capacity), we believe that the extent to which scientific and technical information was used by EPA is typical of many environmental regulatory programs.

Municipal waste combustion (MWC) takes place both with and without energy recovery. Combustion with energy recovery, through generation of steam or electricity during incineration, is termed waste-to-energy (WTE) or resource recovery (RR). Combustion without energy recovery is termed incineration. In the past, both incineration and WTE facilities have been built in the United States. Currently, new MWC facilities almost always include energy recovery.

FRAMEWORK FOR ANALYSIS

There are two points in any regulatory decision-making process where scientific and technical information can be used. The first is during the initial debate about whether a regulatory program is necessary, technically feasible, and cost-effective. The second point is during the development of a rule.

In an ideal world, determining the need for a regulatory program would depend strictly upon analysis of scientific and technical data and information. However, such a world does not exist. Instead, tensions develop between politics and science, both within regulatory agencies and among the different stakeholders in a regulatory outcome. Even once a decision about need has been made, tensions within agencies and among stakeholders continue until a final rule is promulgated, and these tensions may very well persist.

In evaluating whether, and to what extent, EPA used scientific and technical information at these two points in the decision-making process, our analysis focused on the following questions:

• How did the regulatory system respond to scientific and technical understanding?
• What factors contributed to the acceptance (or lack of acceptance) of new information by the risk management and regulatory community?
• What incentives and disincentives existed in the risk management and regulatory system to seek new information?
• What differences, if any, exist in perspectives among federal, state and local levels of government when evaluating new information?

We reviewed documentation available in EPA public dockets for the proposed and final rule and identified additional literature containing scientific and technical information about MWC. In addition, we interviewed representatives from the regulated industry, state and local governments, and EPA.[1] We reached the following conclusions:

• Scientific and technical information was applied in developing particular sections of the proposed and final new source performance standards for municipal waste combustion. However, politics often influenced regulatory decisions, resulting in neglect of available and relevant scientific and technical information. Such information was ignored in determining the need for a federal program specific to municipal waste combustion. Scientific and technical information was dismissed during EPA debates on whether a materials separation requirement should be included in operating permits for individual facilities.

• The perspectives of federal, state, and local governments vary. For example, the stringency of regulatory requirements depends upon the breadth of application in a regulatory program, that is, requirements for a single facility can be more stringent than those applied at a state or federal level with variable environmental conditions and waste management needs.

• Incentives and disincentives for development of new technology are perceived differently among various stakeholders in any regulatory program.

THE MWC INDUSTRY

Municipal waste combustion facilities—with or without energy recovery—are not a recent phenomenon. Nor did the development and implementation of air pollution control technology occur only in response to congressional mandates or agency regulations.

Historical Development

Waste-to-energy and resource recovery facilities have been used as waste management options in the United States for several decades. The nation's first facility constructed with the intent to recover energy began operation in New York City in 1905 (Walsh, 1991). It represented the first attempt at an integrated waste management system, incorporating incineration, recycling, and materials separation at a single facility. Municipal waste was burned in a hand-stoked furnace and energy was recovered with water-tube boilers. The electricity generated was used to light the Williamsburg Bridge. The plant operated for eight years, burning approximately one-fifth of the waste generated in Manhattan and the Bronx while achieving a 60 percent separation and recycling rate. Despite economic success, the facility was closed in 1913 because of maintenance problems (Walsh, 1991). Subsequent incineration facilities built in the 1920s did not include resource recovery and recycling.

In the 1950s, MWC became a recognized waste management tool and source of electricity in Europe. At this time, European vendors were beginning to apply pollution control technology to reduce potentially harmful stack emissions. By the 1960s, MWC reemerged in the United States with the installation of European-developed pollution control technologies. Pollution controls were installed in response to particulate standards promulgated in the late 1960s through air programs within the Department of Health, Education, and Welfare (DHEW) (personal communication with L. Hickman, Solid Waste Association of North America, 1992). Because of a newly perceived solid waste management crisis in the early 1970s, MWC with pollution control regained a broader acceptance by U.S. communities as a waste management option.

As illustrated in Figure 1, use of air pollution control devices predates the development of the 1990 comprehensive new source performance standards. Pollution control devices were first installed in 1957 and became *standard* for *new* facilities by the early 1960s. The first energy-recovery plant to incorporate modern technology was constructed in Chicago, Illinois, in 1970. This facility followed European designs, featured a waterwall furnace for heat recovery and electrostatic precipitators for pollution control, and provided all of its own operating energy. In 1975 a Massachusetts facility sold energy to outside users, initiating the commercial waste-to-energy industry. Also in 1975 an Iowa facility was retrofitted with electrostatic precipitators and fabric filter technology (*Waste Age*, 1992).

Current Industry Status

Figure 2 shows that the use of waste-to-energy as a municipal waste management option has increased dramatically over the past decade. One reason for this increase has been the growing endorsement by government officials of MWC as a legitimate component in national and local waste management plans. For example, in the 1989 EPA report *The Solid Waste Dilemma: Agenda for Action*, MWC was considered a desirable component of the solid waste management hierarchy (EPA, 1989a). The agency's 1993 goal for municipal waste management, as stated in this report, is to reduce the annual volume of waste generated by 25 percent through recycling and source reduction and to incinerate 20 percent, leaving only 55 percent to be landfilled (Porter, 1990). These goals are not mandated by a federal regulatory program.

The volume of waste managed through MWC with energy recovery and the rate at which facilities have been constructed throughout the United States attest to local government's acceptance of this technology as a viable waste management option (Figure 3). Currently, 17 percent of the 196 million tons of municipal waste generated annually is managed at 190 MWC processing and combustion plants (Kiser, 1992). Of these, 142 facilities are waste-to-energy plants with a total capacity of 101,000 tons per day (t/d) (Kiser, 1992). These facilities provide sufficient electricity to meet the needs of 1.3 million homes—equivalent to burning 31 million barrels of oil annually (Kiser and Burton, 1992).

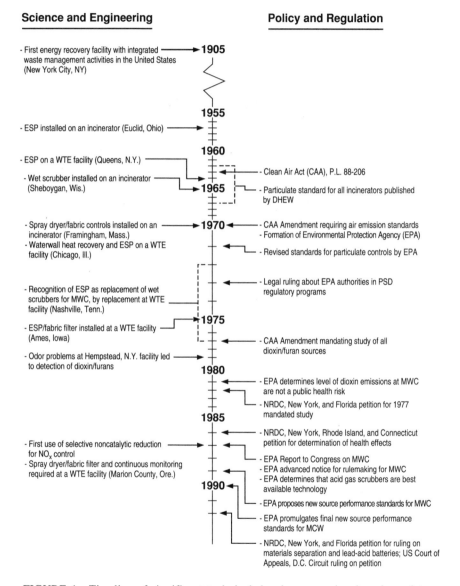

FIGURE 1 Timeline of significant technical, legal, congressional, and regulatory events in combustion of municipal waste.

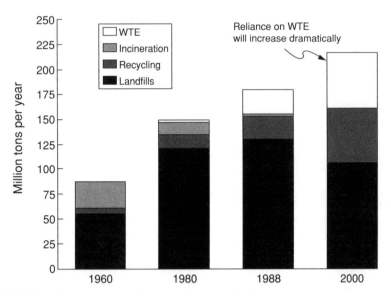

FIGURE 2   Changing trends in managing muncipal waste.

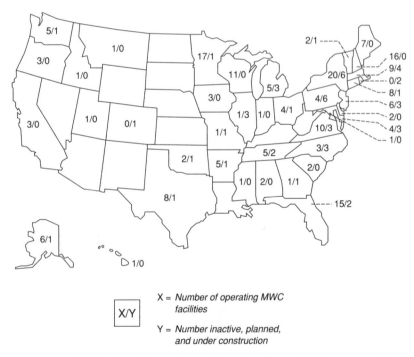

FIGURE 3   Distribution of operating and projected municipal waste combustion facilities.

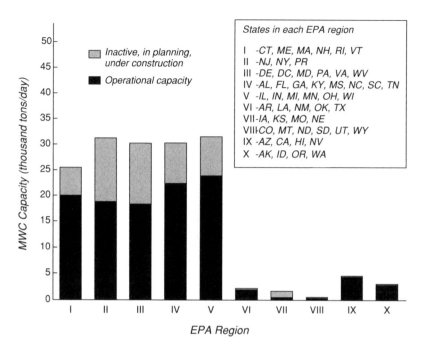

FIGURE 4   Total operating and projected MWC capacity by EPA region.

In addition, projects under construction or being planned could bring total waste volumes managed with MWC to 53 million tons per year (t/y) by the year 2000 (Kiser, 1991). Figure 4 indicates that operational and projected MWC capacity is concentrated in the eastern portion of the United States (EPA Regions I, II, III, and portions of Region IV). For example, 34 percent of operational capacity and 38 percent of projected new capacity are located in EPA Regions I and II alone.

Despite recent increases, however, the level of use in the United States is below that in other nations. Switzerland incinerates 80 percent of its municipal waste, Denmark 60 percent, and the Netherlands 40 percent. Japan incinerates 72 percent of that volume of municipal waste remaining after separation for recycling (Integrated Waste Services Association, 1992a). The increased reliance on MWC in these countries is undoubtedly due to shortages of areas suitable for landfills.

Technology Used Between 1975 and 1989

A major finding of this case study is that between 1975 and 1989, *new* facilities were being constructed with best available pollution control tech-

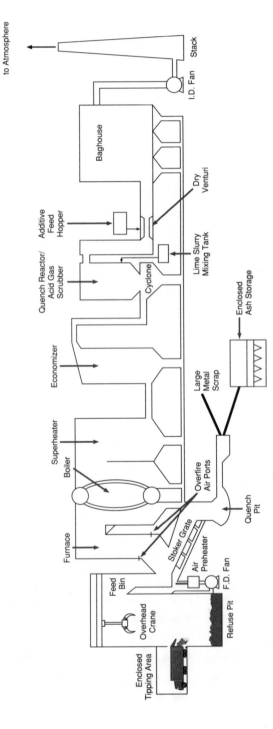

FIGURE 5 Diagram of a mass burn facility. SOURCE: EPA (1987a).

nology and operated with sound combustion practices. These stringent pollution control and good combustion practices were required as part of the CAA Prevention of Significant Deterioration of Air Quality (PSD) program that was managed by state agencies and implemented by local (i.e., county or city) environmental agencies. With few exceptions, these early PSD requirements led to use of the same technologies used by EPA to establish the 1990 new source performance standards.

Combustion Design

Municipal waste combustors consist of three basic types: mass burn, modular, and refuse-derived fuel (EPA, 1989b; National Solid Waste Management Association, 1991). Combustion chambers are similar among these types, with major differences being size of a facility, combustion conditions, and degree of waste processing necessary before combustion. Figure 5 shows the design of a large, mass burn facility.

Mass burn facilities handle mixed waste streams, generally without any precombustion processing other than removal of overly large items and those items included in local source separation programs. Individual units range in capacity from 50 to 1,000 t/d. Facilities can be constructed with more than one combustion unit. Combustion occurs at temperatures ranging from 1,800 to 2,000°F. Typical mass burn technology uses hydraulic rams or pusher grate sections to move waste mechanically onto a grate. Combustion is enhanced by agitating the fuel bed; waterwalls are used to cool the combustion chamber and to recover heat for steam or generation of electricity.

Modular facilities are similar to mass burn plants but are prefabricated and smaller. Individual units range in capacity from 5 to 120 t/d. A modular facility can be constructed with two or more combustion chambers in either of two designs: modular excess-air or modular starved-air. These units are constructed with refractory walls, and most new facilities recover heat with waste-heat boilers.

Refuse-derived fuel facilities use processed waste to ensure a uniform fuel during incineration. Waste processing involves removing materials that can be recycled or are not combustible; remaining wastes are shredded. In these units, shredded waste is generally fed by a stoker onto a moving grate and transported into the combustion chamber. These units range in capacity from 270 to 900 t/d. Virtually all plants are constructed with waterwalls and employ heat recovery systems.

A fourth system under development employs fluidized bed combustion. In these units, waste is burned within a turbulent bed of heated noncombustible material, usually sand or limestone. These units generally burn processed waste, sometimes mixed with other fuels. Design plans generally range in size from 200 to 500 t/d. Heat recovery is generally a component of the design.

Composition of Air Emissions

Several different compounds can be generated and released to the air as a result of combustion activities, for example, operation of motor vehicles, wood-burning stoves, forest fires, and operation of municipal waste combustors. The composition of air emissions formed during combustion depends on the type of material being burned. In an MWC facility *without any air pollution controls*, the type of air emissions generated can vary (DePaul and Crowder, 1988). For example, the following types of compounds may be formed during combustion:

- *Particulate matter* consisting of noncombustible material such as metals, light ash escaping through an exhaust system, or organic material that has not fully been incinerated.
- *Sulfur dioxides* formed from combustion of items such as paper, rubber, wallboard, and grasses.
- *Nitrogen oxides* resulting from the combustion of materials containing nitrogen, for example, yard wastes and textile materials.
- *Carbon monoxide* as a product of incomplete combustion.
- *Hydrogen chloride* may result from combustion of materials containing chlorides.
- *Chlorinated organics* can result from incomplete combustion.

MWC Air Pollution Control Technology

A major difference between facilities constructed after the 1970s and older existing facilities was the extent to which air pollution control was incorporated. In general, local or state environmental agencies have not always required older facilities to retrofit (i.e., to add more efficient air pollution controls). In some instances, agencies perceived retrofits as creating major technical and economic problems that could result in closures of older facilities and disrupt an essential component of the locality's waste management plan. In contrast, new construction designs included requirements for state-of-the-art combustion designs and air pollution controls.

Figure 6 shows the increased use of add-on air pollution control devices over the past two decades. By the time EPA's new source performance standards were proposed for new MWC facilities in 1989, a substantial proportion of new facilities already included the proposed level of air pollution control. The Appendix provides a brief description of the different control technologies.

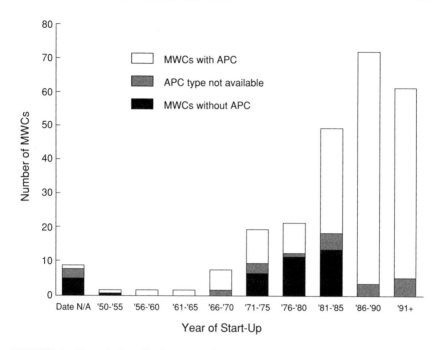

FIGURE 6   Use of air pollution control technology by year of start-up.

Initially, wet scrubber technology was used to control particulates, but this technology soon became obsolete because it could not achieve the particulate standards of the late 1960s and early 1970s. For example, low-energy, wet scrubbers at a Nashville, Tennessee, facility were replaced in the mid-1970s with electrostatic precipitators (ESP) because of the increased removal efficiency and enhanced ability to function at high temperatures (Gaige and Halil, 1992). This change at a large existing commercial facility signaled the end of wet scrubber installations at new units. Later, industry switched to fabric filters and large multifield ESPs for even greater efficiency in pollution control.

In addition to particulates, acid gases (i.e., sulfur oxides and hydrogen chloride) are found in MWC flue gas. Sulfur dioxide (SO_2) removal technology was in its second generation at coal-fired plants during the 1970s; therefore, based on operational information from these plants, this removal technology was easily installed at MWC facilities. For example, in the 1970s, following three PSD permit remands concerned with appropriate acid gas controls, EPA declared that acid gas scrubbers used in conjunction with fabric filters were to be considered "available" control technology in the PSD permit program. By 1987, most new plants were being constructed

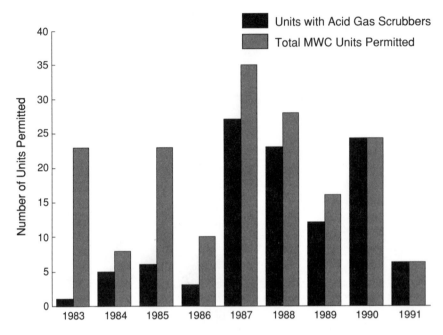

FIGURE 7   Comparison of total MWC units permitted with the number of units constructed with acid gas scrubbers. SOURCE: Gaige and Halil (1992).

with acid gas scrubbers as a result of permit changes for existing facilities (Gaige and Halil, 1992). The rapid use of acid gas scrubbers in the late 1980s is shown in Figure 7.

Emissions of nitrogen oxides (NO_x) can be controlled either by changing combustion conditions or installing add-on controls. Before the 1989 new source performance standards were proposed, combustion practices were used to control NO_x emissions. These consisted of staged combustion with flue gas recirculation and low levels of excess air. This practice can reduce NO_x emissions by 35 percent without add-on controls (Gaige and Halil, 1992).

EPA INVOLVEMENT IN MWC REGULATION

Federal attention to municipal waste combustors began before the creation of EPA. In response to the passage of the first federal statute providing authority to regulate air quality in 1963,[2] officials in the DHEW air program promulgated a particulate standard to be enforced at all incinerators. Federal activity at that time was limited to oversight of state implementation and enforcement of this standard.

Activity at the federal level increased during the 1970s. The 1970 Clean Air Act (CAA) included a provision for prevention of significant deterioration of air quality. Because the statutory language was unclear regarding EPA's authority, controversy arose about the agency's role in PSD activities (Pederson, 1987). Litigation brought against EPA by the Sierra Club in 1973 resulted in a decision that EPA had the authority to require regional air quality standards above national levels and to issue implementing regulations.

Throughout the 1970s and 1980s there was strong public demand to develop stringent air pollution control requirements for MWC within the PSD program. This demand arose in part from legitimate concerns about potential emissions of dioxins and other harmful compounds from uncontrolled MWC air emissions. More recently, this demand has been strengthened by a concern that allowing anything less than the best available control technology (BACT) for MWC facilities could undermine national goals for waste reduction and recycling. For example, public concerns have centered on a perception that "artificially" low combustion costs, resulting from use of less than BACT in PSD permits, would reduce the likelihood that communities could or would be motivated to develop strong waste reduction and recycling programs.

Over the years, Congress responded to public pressure with enactment of several statutes directly affecting MWC:

- 1970 CAA Amendments (P.L. 91-604) mandated new particulate standards;
- 1977 CAA Amendments (P.L. 95-95) required identification of all sources of potentially harmful air emissions;
- 1984 Hazardous and Solid Waste Amendments (HSWA; P.L. 98-616) required evaluation of risks associated with dioxin emissions from MWC; and
- 1990 CAA Amendments (P.L. 101-549) required new source performance standards for MWC.

The regulatory response to these mandates is discussed in the following sections.

Chronology of Legislative, Legal, and Regulatory Activities

EPA has a long history of involvement in MWC, as illustrated in Figure 1. In response to mandates of the 1970 CAA Amendments, EPA issued new source performance standards for particulate emissions for general combustion units larger than 50 t/d and constructed after December 1971. The new standards were issued under the authority of Section 111(b) of the CAA (40 C.F.R. Part 60, Subpart E).

In 1974 the agency issued final regulations for a PSD program. The PSD regulatory program was created as a partnership among federal, state, and local agencies, and became a vehicle for more stringent air pollution control at MWC facilities. The program was delegated to state and local agencies for the implementation and enforcement of air quality standards. State and local agencies were responsible for reviewing proposed new construction or modifications to existing sources. A main component of a PSD permit was a demonstration that BACT would be employed. This demonstration comprised a determination for each new facility that maximum reductions would be achieved, while considering such factors as cost, energy use or development, and nonair environmental impacts.

The 1977 CAA Amendment contained a number of mandates for the agency to address. Among the provisions was a requirement for EPA to investigate all air emission sources of specific compounds, including dioxins and furans. Given the vast number of mandates in the new statute, this investigation did not receive high priority, and EPA resources were allocated to other more pressing mandates (Porter, 1992).

In 1979 EPA's attention was directed to complaints of an odor problem at an MWC plant in Hempstead, New York. Emissions from this facility were monitored, and chlorophenol was detected (Cheremisinoff, 1987). Additional analysis of emissions identified dioxins and furans. Investigations were initiated at five different facilities to determine whether dioxin was commonly emitted from MWC. Based on the results of the investigation, EPA concluded in 1981 that MWC did not pose a threat to human health if dioxin emissions at all MWC facilities did not exceed those identified in the investigation (Cheremisinoff, 1987).

Following announcement of EPA's findings regarding dioxin emissions, environmental groups filed suit against the agency for failing to conduct a full-scale dioxin investigation as mandated in the 1977 Amendments. As part of an out-of-court settlement, EPA agreed to initiate a broader study. Through this subsequent effort, MWC was identified along with other activities as a source of dioxin/furan emissions (personal communication with F. Porter, EPA Office of Air Quality Planning and Standards, 1992). However, the agency did not comment on the level of potential risk that may or may not have been posed by dioxin emissions from MWC compared with the other identified sources, and ignored its previous finding of no health threat. No regulatory action was initiated at this time.

Public concern about dioxin emissions persisted. Congress responded to this concern with additional legislative action in 1984. Section 102 of the HSWA required EPA to prepare a report evaluating risks posed by dioxin emissions from MWC and to identify operating conditions appropriate for controlling these emissions. The EPA subsequently began its investigation and chose to expand its evaluation to include information on other types of MWC air emissions. New regulatory action, however, was limited to promulgation of a more stringent standard for particulate control in 1986. This action was the third change in the particulate standard since the 1963 CAA.

Concerned about the perceived growth in the MWC industry and the lack of a strong federal involvement in the PSD regulatory program, the Natural Resources Defense Council (NRDC) filed a petition requesting development of regulations aimed specifically at controlling air emissions from MWC (Goldstein et al., 1986). The states of New York, Rhode Island, and Connecticut also filed petitions seeking regulations under new source performance standards (NSPS) authority (Abrams et al., 1986). The states were prompted to take this legal action by a belief that the requested new source performance standards (to be developed by EPA) would match an existing program in the New York State (Nosenchuck, 1992).

The report mandated in the HSWA was not delivered to Congress until 1987 (EPA, 1987a). In this report, EPA concluded that control of toxic emissions from MWC could be achieved by the use of good combustion practices, implementation of alkaline scrubbing devices with either electrostatic precipitators or fabric filters, and installation of controls for nitrogen oxides. As EPA acknowledged in the report, the recommended pollution control strategy was representative of those practices and controls already being implemented by state and local governments at many new facilities.

In response to the 1986 petitions filed by NRDC and the three states, EPA announced its intent to develop regulatory requirements for MWC new source performance standards under Section 111 of the CAA. In the advance notice of proposed rulemaking (EPA, 1987b), EPA concluded that, "emissions from MWC may reasonably be anticipated to contribute to the endangerment of public health and welfare." However, the assessment of health risks performed by EPA did not support this conclusion and no other scientific information was cited in the advance notice.

At this time, the agency was also providing guidance to state and local PSD permitting agencies about pollution control and combustion technologies that EPA considered to be BACT (EPA, 1987c). This guidance endorsed those control devices and technologies already being implemented by state and local agencies in constructing and operating new facilities.

In 1989 EPA proposed NSPS regulations for operation of new facilities that matched the 1987 guidelines (EPA, 1989b). In addition to standards for controlling MWC emissions, two other provisions were proposed: a ban on combustion of lead acid batteries and a requirement for materials in municipal waste streams to be separated before combustion of wastes. These two provisions were not included in the final rule in 1991 (EPA, 1991a). These and other changes made between the proposed and final rule are discussed in a later section.

Legal proceedings related to the final rule were brought against the agency in February 1992. The NRDC and the states of New York and Florida challenged EPA's elimination of materials separation requirements and its ban on burning lead-acid batteries in the final rule. A court decision was rendered in July 1992, supporting the agency in its elimination of a materials separation requirement. However, the court supported the petitioners on the issue of lead acid batteries (U.S. Court of Appeals, 1992).

Need for a Regulatory Program

In its evaluation of the need for a federal regulatory program, EPA has an opportunity to evaluate a range of information and data. The types of material include information about potential national impacts of the activity of concern compared with other industrial activities, data pertaining to potential health and environmental risks posed by the activity of concern, and information about alternative options for regulating the activity.

The information available suggests that EPA either limited its development of this type of information or ignored its existence when evaluating the need for a federal MWC regulatory program. Such agency action may have been the result of limited resources; it certainly was a response to political pressures both within and outside the agency. For example, the agency compiled data and information on the likely compounds and emission rates, and conducted two separate risk assessments. However, it did not provide sufficient data or other information that could have presented these rates and potential risks within a national environmental context. In addition, senior officials in the agency apparently chose to disregard the results of the risk assessments.

Projected National Impact

In the advance notice of proposed rulemaking, EPA published estimates of emissions based on projections about the extent of new MWC development (EPA, 1987b). The baseline estimates for "projected facilities" incorporated two assumptions:

• All projected facilities would have efficient particulate removal systems, such as electrostatic precipitators, which were assumed to result in a control efficiency of 99 percent.
• These facilities would be operated using good combustion practices.

Two points must be emphasized to put these assumptions in perspective. First, these *baseline* assumptions were PSD-required operating conditions, which had been implemented at newly constructed facilities since the mid-1980s (see Figures 6 and 7). This fact was noted by EPA in its report

(EPA, 1987a). In addition, some newly constructed (i.e., by 1989) MWC facilities already had been operating with technologies far superior to those noted in the advance notice.

Second, EPA made these estimates based on a projection that 210 new facilities (100 percent increase between 1990 and 1995) would be constructed after promulgation of new source performance standards. The source of this projection is not clearly identified in EPA documents; our analysis suggests that it was an exaggerated number. Considering currently renewed public opposition to incineration, fewer projects may be initiated. Industry data for 1992 supports a conclusion that new development will be much less than the EPA estimate (Kiser, 1992). The number of operating WTE facilities increased only 11 percent between 1990 and 1992. In addition, three facilities are under construction and only 37 are in planning stages. In contrast, 11 MWC facilities have closed since 1990, and 24 proposed projects have been cancelled. While there are more MWC facilities in operation today than in 1989, industry data do not reflect the magnitude of new development projected by EPA.

Having calculated estimates of projected emission volumes, EPA failed to put these estimates in a national context. For example, one component of sound decision making on the need for additional federal air regulations is evaluating the proportion that MWC emissions contribute to total national releases for specific compounds of concern. Information on the relative contributions to national or regional air quality problems would lend scientific credibility to any decision. Using such information, EPA could have ranked the need for additional federal regulations for MWC among the range of environmental issues facing the agency. In a climate of limited resources, EPA (and Congress) must be able to identify priorities that would provide the greatest health and environmental protection in a cost-effective manner.

A preliminary comparison of MWC emissions with other industrial releases suggests that failure to conduct such an analysis hindered a sound decision about the need for an MWC federal program. Industry estimates of new development are only about one-fourth of those projected by EPA. For our comparison, we conservatively assumed that there might be a 50 percent increase in new facilities by 1995. Using this projection and EPA emission rates, the following examples illustrate the impact of MWC compared with other industrial activities. For example:

• MWC air emissions of cadmium could represent 7 percent of the total amount released by the metals industry;
• MWC polychlorinated biphenyl (PCB) emissions might be equivalent to 6 percent of total PCB releases by the electrical industry;
• Chromium releases from MWC could represent 1 percent of total chromium releases by the electrical industry;

• Chlorobenzene emissions from MWC might correspond to 0.1 percent of reported releases by the chemical industry; and
• SO_2 emissions could represent less than 0.1 percent of these emissions from utilities (Reisch, 1992).

Using current mercury emission data, MWC is estimated to contribute less than 1 percent of total mercury released to the environment (Kiser and Sussman, 1991). This limited comparison suggests that the basis on which new development is projected becomes a critical factor in analyzing regulatory needs. If EPA had applied the results of a comprehensive, comparative analysis of projected air emissions, its decision about a federal regulatory need might have been much different. In the absence of reliable and verifiable data, EPA staff had no ability to balance diverse political pressures.

The expected reductions in air emissions associated with implementation of the proposed standards (EPA, 1989b) had little scientific credibility. For example, estimates of projected reductions were based on expected achievements by mid-1990, assuming that all 210 new facilities would be constructed and operating. In developing these estimated reductions, EPA did not acknowledge that facilities constructed since the mid-1980s *already included these proposed technologies.*

Two factors influence the agency's effectiveness in making critical national decisions. The first concerns limited resources and timing of necessary decisions. Resource and time constraints hinder EPA's ability to develop sufficient data with which to make decisions; thus, its decision-making effectiveness is weakened and general acceptance of final decisions is jeopardized. Second, rarely does the agency assume a necessary leadership role in making tough decisions that are counter to public fears, even when the agency's own information indicates that these fears are without basis. The EPA could have insisted that an additional program was unnecessary because the available information indicated that MWC emissions posed no national threat and because state and local agencies were addressing local potential risks.

If appropriate data had been developed early, it might have affected the congressional debate in the early 1980s. Knowledge about the magnitude of perceived threats and the potential for enhanced protection, evaluated within a national context, also might have balanced political pressures for federal action. With sound data on MWC industrial development, evidence of limited risks to public health, and increasing local community acceptance of these facilities, the agency could have placed itself in a leadership role in the debate about the need for an additional federal program. Instead, EPA placed itself in the unfortunate position of being only one of several stakeholders, all of which had information of limited scientific or technical credibility.

Evaluation of Risks

A second type of information necessary to evaluate the need for a federal regulatory program concerns potential risks to human health and the environment posed by exposure to MWC emissions. As part of its 1987 Report to Congress, EPA conducted risk assessments of MWC emissions (EPA, 1987d). The results suggested that individual risks (i.e., those posed by a single facility to residents of a host community) were not above the then-informal, national acceptable risk range.[3] The agency evaluated potential risks under two exposure conditions:

• Projected facilities would use electrostatic precipitators to control particulate emissions and implement good combustion practices to minimize emissions of other constituents; this was designated the baseline scenario.

• Projected facilities would use dry alkaline scrubbers in conjunction with particulate control technology and good combustion practices equivalent to the baseline condition; this was designated as the controlled scenario, which was being installed at many new facilities in 1987.

The EPA estimated an individual lifetime cancer risk from direct inhalation for the baseline scenario as a range of 1 in 100,000 (10^{-5}) to 1 in 10,000 (10^{-4}) for organic compounds and 1 in 1,000,000 (10^{-6}) for metals; these risks are within acceptable ranges (i.e., 10^{-6} to 10^{-4}). In this assessment, EPA indicated that the primary contributor to any risk associated with organics was the level of estimated dioxin and dibenzofurans emissions. Actual data obtained during EPA's 1981 investigation of five facilities were counter to these new estimates.

The discrepancy between estimated emission levels and actual monitoring data can be explained. For example, the assumptions and values for exposure variables used by the agency in this assessment are extremely conservative (Roffman and Roffman, 1991). The EPA assumed a stack height that is unusually low and might not have been acceptable under 1989 state and local PSD permits. In addition, estimates of emission rates for selected constituents may not have reflected the level of pollution control also being required by state and local governments at that time. Therefore, given the level of conservatism in EPA exposure assumptions, this risk estimate is likely to overestimate any real risk. The EPA indicated in its advance notice of proposed rulemaking that "[i]t is not likely that the true risks would be higher than the estimated risk, and *they may be considerably lower*" [emphasis added] (EPA, 1987b, p. 25404). Because the agency's estimates of potential risk are within the acceptable risk range and because EPA believed that actual risks would be lower than the estimate, there is no justification for its view that regulation was warranted.

During the course of preparing proposed standards for new MWCs, EPA conducted a second risk assessment (Morrison, 1989). In this analysis, emission data were collected by the agency directly at newly operating facilities. The baseline and control scenarios, as well as the assessment methodology, were the same as in the previous study, but the quality of data was better in the 1989 analysis. The EPA indicated that emission data used in 1989 were "generated using prescribed testing, analytical and Q/A [quality/assurance] procedures" (Morrison, 1989, p. 2). Such standard collection and analysis precautions were apparently not followed in the 1987 effort.

Results of this second analysis estimated an overall worst-case individual lifetime cancer risk of 7 in 1,000,000 (7×10^{-6}) for the baseline scenario. Estimates of noncarcinogenic hazards associated with air emissions for hydrogen chloride, lead, and mercury were below health parameters enforced in 1989.[4] Therefore, facilities with pollution control technology as defined in the baseline scenario do not pose unacceptable carcinogenic risks or noncarcinogenic hazards. Agency officials chose to ignore these studies.

More recent publications discussing risks associated with MWC emissions support these 1989 estimates, suggesting that risks from exposure to MWC emissions are minimal for host communities. For example, dioxin levels (the reason for much of the public concern) from new facilities have been measured at less than 1 percent of estimated background intake. Concentrations of inorganic and metal emissions are similar or lower than those detected as background levels in rural environments (Greim, 1990).

Options for EPA Involvement in Air Regulatory Programs

In 1985 EPA unveiled a new National Air Toxics Strategy to address regulatory needs (EPA, 1985). In this strategy, the scope of any national regulatory program was to target those "sources of complex toxic emissions that appear to account for a significant portion of the controllable health risk" (EPA, 1985). In this strategy report, the agency identified those sources of air emissions meriting priority attention for regulatory development. MWC facilities were not mentioned.

In testimony to Congress, EPA Administrator Lee Thomas emphasized a desire in EPA to make choices between the need for developing new national programs or for greater cooperation with state air-toxics control programs (Thomas, 1985). A decision for greater cooperation with state programs could be based on the recognition that the potential risks posed are local (or regional) in nature and do not represent a major national public health risk. The EPA appeared committed to assisting states in addressing these more local problems.

Therefore, at the time that decisions about controlling MWC emissions were being debated, the agency had two options. The EPA could develop a new federal regulatory program, or it could enhance EPA cooperation and assistance, both financial and technical, to state regulatory programs. Information available to EPA at that time suggests that the latter option could have been preferable. For example:

• MWC facilities were (and still are) "clustered" in the northeastern section of the nation. This technology was being used as an alternative at those locations where landfill capacity was most severely limited.

• New facilities constructed in the mid-1980s generally included state-of-the-art air pollution controls in response to state and local PSD permit requirements.

• Emission levels estimated by EPA under its baseline scenario represented a small percentage of national emissions for specific constituents and were of a much lower magnitude than those emissions from other industries. Therefore, these facilities did not pose a major national public health threat.

• The EPA's own risk estimates for projected facilities indicated that risks associated with these new facilities were within the range of acceptable risks.

At this point, many state and local environmental agencies were developing stringent requirements for new facilities. A more efficient process that could be supported with existing scientific and technical information might have been to leave the regulatory response at the state level with continued assistance from EPA. Using this information, EPA could have justifiably concluded that the most cost-effective means of regulating these facilities would be to strengthen state and local environmental programs. Constrained federal resources could have been directed at developing regulations for more significant national pollution problems.

Choice of Regulatory Authority

Two problems were presented to the agency. One was the fact that its own data suggested that there were no unacceptable health risks associated with MWC emissions. The second problem was the public pressure for federal regulations beyond those already included in the PSD permit program. In addition, there was evidence that once Congress addressed amendments to the CAA, it would mandate development of new source performance standards for MWC facilities. Thus, ignoring the scientific and technical data at hand, EPA proceeded to seek a legislative mechanism that did not depend on health- or risk-based standards as the foundation for its regulatory program.

Two authorities exist for development of air emission standards in the

New Source Performance Section of the Clean Air Act: Section 111 and Section 112. Section 111 specifies that standards of performance for new stationary sources are to be developed based on best demonstrated technology. Standard of performance is defined in the statute as (42 U.S.C. §7411, Sec. 111 (a)(1)):

> . . . reflecting the degree of emission limitation achievable through the application of the best system of emission reduction which (taking into account the cost of achieving such reduction and any nonair quality health and environmental impact and energy requirements) the Administrator determines has been adequately demonstrated.

In contrast, Section 112 requires development of compound-specific or category-specific, health-based standards. Constituent concentrations developed for these standards must be related to scientifically developed threshold levels representing no adverse health effects.

The 1986 petition by NRDC and three states specifically requested that EPA use the Section 112 authority for regulating MWC emissions (Abrams et al., 1986; Goldstein et al., 1986). However, both EPA and industry preferred to use Section 111. The EPA believed that:

> . . . in view of the broad range of health effects and the multiple constituents of MWC emissions, the use of section 111(b) and 111(d) constitutes the most appropriate, comprehensive regulatory strategy for control of these emissions (EPA, 1987b, p. 25399).

Industry concurred. Thus, by selecting Section 111, the agency could use a combination of design, equipment, and work practice or operational performance as the standards of performance. This choice removed the scientific and technical dilemma for the agency, that is, the inability to justify a regulatory program based on potential public health risks. Standards developed under Section 111 do not have to be supported by risk analyses, nor are they required to have a health basis.

Factors Leading to Federal Involvement

Since the early 1970s, MWC has been viewed by many states and local governments as a viable alternative to landfilling. However, because of public concern about potential risks associated with these facilities, local regulatory agencies have been motivated to implement state-of-the-art technologies as a means of gaining community acceptance of proposed facilities. Unless these officials are able to show that the best available air pollution controls will be installed, their communities do not approve construction (personal communication with D. Gatton, U.S. Conference of Mayors,

1992). Consequently, most new facilities being permitted before development of an EPA regulatory program employed the type of technologies discussed earlier.

Our analysis indicates that internal and external political pressures were the single most important factor influencing EPA's decision about a federal regulatory program. The agency ignored and disregarded available information that was counter to these political pressures. For example:

• The EPA suggested that the number of new projects and amount of waste to be managed at these facilities could be expected to grow significantly as landfill capacity declined. While some projection was acknowledged by industry, the agency discounted available information that projected development would not be uniform across the United States. The economic costs of MWC cannot be borne by all communities or regions. Therefore, the expected pattern of development would be dependent on the availability of landfill capacity and cost-effective recycling programs as a means of reducing disposal volumes. Available information indicated that MWC development was highly regional and much more local than suggested by EPA.

• The EPA provided no basis to discount the expectation that state and local regulatory agencies would continue requiring stringent PSD permit regulations. In addition, there is no reason to expect that EPA oversight of these requirements and resulting permits could not continue and would not be effective. The agency did not follow its own 1985 National Air Toxics Strategy, which could have increased financial and technical assistance to state environmental programs.

• EPA estimates of increases in overall MWC air emissions, without a new federal regulatory program, were not supported by any information available in the mid- to late-1980s. In fact, these estimates were based on less efficient technology and less rigorous operating practices than were already being required by state and local regulatory officials at new facilities. Also, the agency did not attempt to place estimates within a national context as a means of determining whether they were a major contributor to public health and environmental risks.

In addition, a new administration was taking office at the time EPA was debating its decision. This new administration had emphasized a commitment to the environment. Therefore, MWC became an easy and very visible target to demonstrate this federal commitment, regardless of the lack of scientific support for a need to develop another federal regulatory program.

Too often, pressures from public interest groups and Congress overwhelm any attempt by the agency to develop regulatory programs that incorporate sound scientific and technical information. Unfortunately, there are examples in other regulatory programs where political pressure negated an EPA decision to develop regulations heavily weighted toward scientific and technical information. In such situations, conflict with Congress appears inevitable[5] and EPA is prevented from taking a leadership position armed with sound scientific and technical information.

NEW SOURCE PERFORMANCE STANDARDS

A second stage at which scientific and technical information can be applied in a regulatory context is during development of a proposed and final rule. For the most part, EPA relied on these types of information in the new source performance standards for MWC emissions for new facilities. The following sections provide examples of the information used and various points where new information emerged.

The purpose of the NSPS was clearly stated in the preamble of the final rule:

> The intended effect of these standards is to require new, modified, and reconstructed MWCs to control emissions to the level achievable by the best demonstrated system of continuous emission reduction, considering costs, nonair quality health and environmental impacts and energy requirements (EPA, 1991a, p. 5488).

In developing these standards, EPA must determine which control technology or technologies can be considered as best demonstrated. Therefore, it must gather control data for each pollutant, or class of pollutants, of concern by sponsoring new research and by requesting data from operating facilities. Section 111 emphasizes that the control achieved by these technologies must be obtainable in a consistent manner across all facilities.

Numerical standards presented in the NSPS rule are based on the capabilities of specific technologies. Although specific control technologies are identified as the basis of particular standards, there is no requirement that these technologies must be used. Any control technology that will achieve the numerical standards can be implemented (EPA, 1991a).

Provisions of Section 111 make use of scientific and technical data in a rather straightforward manner. As indicated previously, the definition of "standard of performance" in Section 111 is the best demonstrated technology (BDT) available to control emissions of concern. However, even with an emphasis on a technology-based standard, political pressure can affect this stage of regulatory development rather rapidly and forcefully, as illustrated in the following discussion of the materials separation requirement. As noted there, politics worked against using available information in the proposed rule, but was the driving force (in support of available information) in changing the provision in the final rule.

Development of Specific Standards

The EPA did, indeed, use scientific and technical information as the basis for the majority of these standards. In addition, changes from the proposal to the final version were supported by new data and information acquired through the public comment process. This information included new technical performance data generated at and provided by operating facilities, results of economic analyses received during public comment, and questions about the validity of data used by EPA in developing the proposed rule. Often such questions of validity prompted the agency to reevaluate the relevance or credibility of information in its data base.

Table 1 lists the changes made between the proposed and final rule. The extent to which technical and scientific information has been used is illustrated below for good combustion practices and emission controls for particulates, metals, acid gases, organics, and nitrogen oxides.

TABLE 1   Comparison of Proposed and Final NSPS for MWC

Standard	Proposed Rule	Final Rule
Applicability	Separate requirements for <250 and >250 t/d	Only >250 t/d
Good combustion practices (GCP)	50, 100 or 150 ppmv CO, depending on technology (4-hr averaging time) 100% maximum MWC unit load determined during dioxin/furan test 230°C at PM control device inlet	Same limits as proposed, but some averaging times revised to 24 hours 110% maximum MWC unit load 17°C above maximum temperature at PM control device inlet
Metals (PM)	34 mg/m^3 at 7% O_2 10% opacity	34 mg/m^3 at 7% O_2 10% opacity (6-min average)
Acid gases	80% reduction or 30 ppmv SO_2 95% reduction or 25 ppmv HCl	85% reduction or 30 ppmv SO_2, at 7% O_2 (24-hr block geometric mean) 95% reduction or 25 ppmv HCl, at 7% O_2 (annual stack test)
Organics	5 to 30 ng/m^3 total dioxins and furans	30 ng/m^3 total dioxins and furans
Nitrogen oxides	120 to 200 ppmv	180 ppmv at 7% O_2 (24-hr block average)

Good Combustion Practices

The term good combustion practices (GCP) refers to a set of operating principles and procedures that ensure optimal operation of MWC facilities. This optimization maximizes complete combustion of organics and minimizes formation of dioxins and furans. Components of GCP include identification of (EPA, 1991b):

• Limits for carbon monoxide (CO) emission levels;
• Site-specific maximum inlet temperature for particulate control devices, demonstrated during a dioxin/furan performance test;
• Maximum load level demonstrated during a dioxin/furan performance test; and
• Certification for supervisor and operator training and use of training manuals for other personnel.

Changes made in these components and published in the final rule reflect the incorporation of new information provided through public comment. For example, the proposal required monitoring CO levels to verify correct implementation of GCP. The proposed CO limits and averaging times were based on data from existing facilities (EPA, 1991a), but data from newer facilities indicated that lower limits, as eventually promulgated in the final rule, could be achieved (EPA, 1991b).

Similarly, new data were the basis for changes in maximum temperatures at particulate control device inlets. The single temperature proposed for particulate control devices was changed to a site-specific maximum (EPA, 1991a,b). Several alternative temperature limits were recommended by industry during the comment period in an effort to improve metals emission control. However, EPA indicated that the purpose of this standard was to prevent formation of dioxins/furans rather than to control metals emissions. Because temperature variability influences the generation of dioxins/furans, the final rule established a maximum variance (e.g., + 17°F) for these temperatures based on new industry information (EPA, 1991b).

The EPA changed the performance standard for maximum steam load in response to new data about technological feasibility. The objective of this standard is to limit flowrates of flue gas, maximize control of particulate emissions, and reduce formation of organics. The agency enforces a maximum steam load requirement because flue gas flowrates, themselves, cannot be measured accurately. Equipment vendors are developing systems to measure flue gas flowrates; once one is validated as best demonstrated by EPA, actual measurement of flowrates can be incorporated into the standard (EPA, 1991a). This final standard clearly illustrates EPA's acceptance of limitations in both technical achievability and monitoring capabilities when developing regulations. Because Section 111 requires EPA to periodically review all standards, there will be future opportunities to revise the basis of new source performance standards as new technology becomes available.

Particulate and Metal Controls

The standard for particulate and metal controls was not changed in the final rule. However, some commenters felt that the proposed standard was too stringent or not achievable. In response to these comments, EPA sought and reviewed data from 10 plants with state-of-the-art control devices. These data indicated that average particulate emissions could be controlled to below 23 milligrams per dry standard cubic meter (mg/dscm). One set of data, however, indicated emissions of 32 mg/dscm. The best demonstrated technology is defined as that technology which can consistently achieve certain emission concentrations. Therefore, EPA selected a final standard of 34 mg/dscm (EPA, 1991a).

Acid Gas Controls

Acid gas standards were the source of some controversy during public comment; concerns were directed at questions about what could be considered an appropriate designation of best demonstrated technology. Some of those contesting the proposed standard suggested that more long-term data were needed and that it would be preferable to establish design standards (e.g., flue gas temperature entering the control equipment or stoichiometric ratios) rather than numerical limits (Walsh, 1990). However, EPA maintained that its data were sufficient to demonstrate that acid gas limits could be met.

Between publication of the proposed rule and promulgation of the final rule, EPA reviewed additional data supplied by commenters and changed its standard for acid gas emissions. The SO_2 standard was revised as noted in Table 1 (EPA, 1991a). Considerable new information was put at EPA's disposal—some indicating that more stringent standards can be achieved and other data suggesting that greater stringency was not possible. Information about marginal costs, increased volumes of lime necessary to achieve greater control, and trade-offs between incremental acid gas reductions and increased landfilling of lime wastes was provided to EPA.

To resolve the controversy, the agency evaluated SO_2 emissions data from a new facility equipped with spray dryer/fabric filter controls and determined that an 80 percent reduction could be achieved consistently and continuously. In this case, a change in a final standard was based on statistical analyses of data from an actual facility (EPA, 1991a,b).

Although some commenters believed that the standard for hydrochloric acid (HCl) was too stringent, EPA did not agree. Tests from a number of facilities with spray dryer/fabric filter controls indicated that 95 percent reduction could be consistently achieved (EPA, 1991a). The agency determined that HCl control technology, tested annually, was sufficiently developed, and therefore demonstrated, to meet the standard consistently (EPA, 1991b).

There also was some controversy associated with monitoring for acid gases. Sulfur dioxide emissions can be monitored continuously; however, continuous HCl monitors are not yet "demonstrated." Available data indicated that the HCl standard (95 percent reduction) is achieved when SO_2 is reduced by 80 percent. Therefore, EPA determined that continuous HCl monitoring represented a redundant requirement. The agency supported this determination with statistical correlations between HCl and SO_2 removal (EPA, 1991b).

Rather than a numerical concentration limit, EPA selected a percent reduction performance standard because it could be demonstrated as most accurate and representative of performance for acid gas control systems. Requiring intermittent, high emissions to be reduced to a specific numerical level could have unnecessarily increased the complexity and costs of control systems (EPA, 1991a). The agency applied information on technological capability in setting these standards.

Organic Controls

The EPA proposed to control emissions of organics by using dioxin/furan emissions as a surrogate measure of formation of organics. The proposed rule indicated a concentration range and stipulated that a single limit would be promulgated in the final rule (EPA, 1989b). In finalizing this standard, EPA analyzed data generated at ten facilities. Emissions were below 10 nanograms (ng)/dscm at eight facilities, but concentrations of up to 29 ng/dscm were observed at two (EPA, 1991a). The final standard for MWC organic emissions is 30 ng/dscm, a level that EPA determined could be consistently achieved.

This standard generated some controversy during public comment. First, there was concern whether dioxins and furans are suitable surrogates for organic emissions. Second, the agency was proposing to measure *total* concentrations of dioxins and furans, which does not account for differences in toxicity among the different congeners and isomers. Third, the methodology proposed for analyzing dioxin/furan concentrations was considered inadequate for the task.

Suitability as Surrogates. Measures of dioxins/furans were selected by EPA as a surrogate for organic emissions for several reasons:

• Results of the 1987 risk assessment indicated that potential carcino-genic risk was predominantly due to the presence of dioxins/furans and no other organics (EPA, 1987d).

• Carbon monoxide (CO) levels are indicators of incomplete combus-tion and the potential for formation of organics and dioxins/furans; rank order correlation of data from the 1987 MWC study indicated a positive relationship between these variables; thus EPA believes that CO limits, in addition to limits on dioxin/furan, would ensure complete combustion and minimal formation of organics (EPA, 1991a,b).

• Data indicated that CO concentrations greater than 200 parts per million by volume (ppmv) are associated with uncontrolled formation of dioxins/furans (EPA, 1989b); because dioxin/furan emissions are tested only during an annual compliance test, CO limits serve as a continuous basis to determine if there are excessive dioxin/furan emissions.

Total Dioxin Measurements. In the proposed rule, the agency noted that although it was theoretically possible to measure all components of flue gases, it would be very burdensome, expensive, and impractical to perform such analyses repeatedly (EPA, 1989b). Therefore, the agency relied on total dioxin levels as the best method of control.

The International Toxicity Equivalency Factors (I-TEF) is an example of methods used to estimate relative toxicity of various dioxin and furan congeners and isomers (EPA, 1989c). Using the I-TEF concept, linear re-gression of emissions data obtained in the MWC study indicate that "mass emission measures based on either TCDD concentration or PCDD/PCDF concentration can be used as surrogates for toxic equivalency measures in analyses of PCDD and PCDF emissions" (EPA, 1987b). The EPA used these data to justify application of total dioxin measurements. In addition, EPA based its decision on information that there was no demonstrated method to control emissions of various isomers selectively. The agency also indi-cated that allowances would be necessary for differences in TEF values used by different agencies and for emerging new information (EPA, 1991a). In addition, EPA considered it inappropriate to use toxic equivalency as a basis for a standard developed under Section 111:

> . . . emission limits for total dioxin/furans reflect the achievable perfor-mance levels of specific types of control technologies, and are not derived from any target levels of health risks (EPA, 1991a, p. 5504).

Analytical Methodology. Comments on the proposed rule expressed con-cern about EPA's choice of an analytical method to detect dioxins. The method was believed to be inaccurate and result in overly high concentra-tions for particular congeners. Without evaluating new information, EPA maintained its provision to require Analytical Method 23 and asserted that it was adequate to determine compliance with the standard (EPA, 1991b).

Nitrogen Oxides

The agency based its decision to promulgate a standard for NO_X emissions on information about the availability of technology, the projected rapid growth of new facilities, and the fact that control costs for this pollutant are not unreasonable. Furthermore, Section 129 of the 1990 CAA Amendments requires EPA to regulate NO_X emissions at new and existing facilities. Therefore, because BDT exists for NO_X control, limits were established (EPA, 1991b).

The EPA initially proposed a range for NO_X emissions. Following the proposal, the agency evaluated extensive continuous monitoring data from a grate-fired mass burn facility with Thermal DeNOx controls. The EPA concluded that a single numerical limit could be consistently achieved. Data from 35 other facilities, representing different types of combustion technology, supported this conclusion (EPA, 1991b). The EPA also believed that some designs might even meet the standard without add-on control (e.g., water-cooled rotary mass burn MWC).

A controversy emerged during public comment about whether selective noncatalytic reduction (SNCR) is a demonstrated, developing, or experimental technology. At that time, the only NO_X controls in use were at three facilities, and opponents contended that data from three facilities did not constitute best demonstrated technology. The agency responded that SNCR had been in use since 1987 and that available data were sufficient to establish this technology as BDT (EPA, 1991b).

Issues about potential toxicity of a detached ammonium chloride plume observed when SNCR is used were raised by commenters. The EPA does not consider ammonium chloride emissions to be environmentally significant; in addition, it contends that formation can be controlled with proper operation.

Materials Separation

In contrast to the extensive application of scientific and technical data used by EPA in formulating the other provisions of the proposed and final NSPS, the proposed materials separation requirement was developed in the absence of supporting scientific or technical information. Inclusion of this requirement was driven by internal political considerations. The debate concerning a materials separation requirement is discussed at length here because it is the most dramatic example of a conflict between political desires and supporting scientific, technical, and economic information.

Rationale for Proposed Requirement

The materials separation provision mandated separation and removal of certain materials from municipal waste streams before combustion. Under the proposed rule, an MWC facility's NSPS operating permit would include a specific requirement that 25 percent (by weight) of waste received at the permitted facility must be diverted from the combustion process. If at any time during the permit life, this diversion requirement was not met, the facility would not be in compliance with the permit. The EPA's justification for this provision followed two lines of reasoning (EPA, 1989b). The first line incorporated a belief that:

• Most separated materials could be recycled, thereby preserving natural resources, reducing volumes of waste to be landfilled, and potentially providing communities with income from these recycling activities to partially offset MWC operational costs; and
• Reducing the amount of waste combusted would reduce both total MWC emissions and those of specific constituents.

The second line of reasoning indicated that certain environmental benefits could result from implementation of the provision (EPA, 1989b). For example, separating paper and paperboard would:

• Reduce specific metal emissions because lead-based inks and mercury-based fungicides would not be burned;
• Lower carbon dioxide emissions because less material would be available for combustion;
• Preserve forestry resources by recycling paper products;
• Reduce HCl emissions by removing bleached paper; and
• Reduce SO_2 emissions by removing combustion of sulfur-laden paperboard.

Furthermore, EPA suggested that reduced combustion of paper would result in the downsizing of MWC units because lower volumes of waste would require incineration (EPA, 1989b). However, the agency provided no information to support the concept that downsizing of facilities would result in an environmental benefit.

Scientific and Technical Data

The 1990 report *Municipal Waste Combustion: Background Information for Materials Separation* presented EPA's evaluation of data to support its claims about reduced emissions associated with materials separation (EPA, 1990). Emissions monitoring data for several constituents were presented, including data for total metals, mercury, lead, dioxins and furans, paper, and yard waste. These data were derived from three mass burn facilities located in Nashville, Tennessee; Salem, Virginia; and Gallatin, Tennessee. Emission samples were collected upstream from pollution control devices to represent uncontrolled emissions and did not reflect concentrations or quantities emitted from the stacks. Data were analyzed on the basis of the total amount of waste processed, not the amount burned. These data did not demonstrate that a *consistent* reduction in MWC emissions would be achieved.

Table 2 shows the range of variability in EPA's data for compound-specific emissions. For example, lead concentrations following materials separation varied from a decrease in emissions of 52 percent to an increase of 5 percent over baseline emissions. More startling, carbon monoxide levels ranged from a decline of 65 percent to an increase of 468 percent over baseline emissions (EPA, 1990).

At the time of proposal, EPA acknowledged this lack of consistency in the data concerning emission reductions as well as the problems inherent in the basic study design. For example, data representing emissions when waste streams were not separated and those representing separated wasted streams were collected months apart at a single facility without any control for variations in the types of wastes collected. In addition, data for nonseparated and separated wastes were not always collected from the same facility. Such collection practices undoubtedly contributed significantly to observed variations in emissions.

The result of this data collection effort was that EPA could not provide supporting evidence for its theory about the impact of materials separation on MWC emissions (EPA, 1990). For example:

TABLE 2   Comparison of Emission Changes Associated with Materials Separation

Pollutant	Range of Change (percent)	Nashville, Tenn.	Salem, Va.	Gallatin, Tenn.
Particulates	+20	↑		
Arsenic	−70 - −5	↓	↓	↓
Cadmium	−73 - −22	↓	↓	↓
Chromium	−63 - +68	↑	↑	↓
Lead	−52 - +5	↓	↑	↓
Mercury	−71 - +260	↓	↑	
Carbon monoxide	−65 - +468	↓	↑	↓
Nitrogen oxides	−42 - +8	↓	↓	↓
Sulfur dioxide	−1 - +11	↑	↑	↓
Hydrogen chloride	−79 - +17	↓	↓	↑
Total hydrocarbons	−74 - +214	↓	↑	↓

SOURCE: Adapted from EPA (1990).

• Levels of dioxins/furans appeared to be unaffected by the removal of polyvinyl chloride from combusted waste; HCl emission levels appeared to be correlated with this removal.

• The EPA acknowledged that "there are no data to demonstrate what quantity of mercury, cadmium, or other metal emissions from MWCs are due specifically to the combustion of batteries." The actual impact of removing metals before combustion could be quite small, given that all new facilities must have highly efficient air pollution controls.

• Removal of over half of the total lead content from waste, on average, did not provide consistent net reductions in lead emissions.

• The agency acknowledged that "[n]o data are available on the positive or negative effects of paper separation on MWC emissions."

• The EPA could only *speculate* that removal of yard waste would result in reduced NO_x emissions; the basis of this speculation was that higher NO_x concentrations are observed in summer, when the waste stream contains more yard waste.

Nevertheless, EPA ignored the lack of consistent evidence to support its theory on the environmental benefits of materials separation and persisted in including a materials separation requirement in the proposed rule (EPA, 1989b).

Political Debate Concerning Materials Separation

Although all stakeholders in the final rule for new source performance standards have been extremely supportive of recycling programs, many of them opposed requiring materials separation as a component of MWC permits. These latter included state and local governments, the MWC industry, and particular offices within the executive branch. Only environmental groups expressed strong support for continued inclusion of this provision. The comments made by these groups are summarized in the following sections to illustrate the different perspectives of all stakeholders and the type of new information uncovered during the public comment period.

The "influencing factors" in this debate were concerns about legal authority, lack of evidence that the requirement could be implemented, and lack of evidence that emissions could be consistently reduced. The core argument against the provision, however, was a concern that a better mechanism for initiating recycling programs existed, that is, the regulatory program for new source performance standards—which is aimed at preventing air pollution—was not an appropriate vehicle to mandate community recycling programs. The EPA never fully explored any other mechanisms and made little, if any, effort to obtain information about actual coexistence of MWC facilities and community recycling programs.

A recent survey by the Integrated Waste Services Association illustrates the type of information that should have contributed to EPA's decision about materials separation (IWSA, 1992b). This survey indicated that recycling programs and MWC combustion are compatible waste management activities and coexist without regulatory requirements being included in MWC operating permits. For example, in those counties and cities with operating MWC facilities, recycling rates generally exceed the proposed 25 percent goal. These recycling rates range from 24 to 46 percent. This industry survey suggests that communities are able to develop successful recycling programs without mandating materials separation as part of a NSPS permit. The survey does not address the impact of these recycling programs on emissions from facilities.

State and Local Governments. In comments submitted to the Public Docket, state and local governments contended that the CAA was not an appropriate authority under which to require materials separation and recycling programs. Most commenters expressed a preference that such provisions were best handled within authorities of the Resource Conservation and Recovery Act (RCRA) or by a presidential initiative (EPA, 1991b).

While state and local officials generally supported recycling as a major component in the solid waste management hierarchy, they objected to having this requirement as part of MWC operating permits. The materials separation provision could result in numerous difficulties for local communities, such as certifying that 25 percent of the waste stream had been separated and problems associated with maintaining consistent recyclable waste volumes. In the latter case, facilities would have to cease operation until the target recycling rate was possible.

A major factor influencing achievability of specific recycling rates is the lack of or uncertain markets for recyclable materials. In the absence of markets, local governments and operators of a facility would be faced with the need for long-term stockpiling of separated materials. Although EPA proposed a waiver provision for burning nonrecycled materials should markets not be available at particular times, local officials emphasized the cumbersome nature of the waiver process.

In addition, contracts between local governments and operators of these facilities generally cannot be adjusted readily. These contracts usually include a cost penalty placed on local governments when specified combustion volumes are not met. Therefore, any delays in obtaining these waivers would place financial burdens on local governments. The contracting process for operation of MWC facilities is not conducive to a materials separation requirement as part of an NSPS permit (Curling, 1990; Martineau, 1990a; National Association of Counties, 1990).

Echoing many of these local issues, state officials also emphasized their concerns about quantifying and certifying materials separation rates. They questioned whether there are methods that can provide consistent definitions for quantifying baseline levels and subsequent reductions. If these methods are not available, documentation of compliance with a materials separation standard is not possible (Nosenchuck and Bruckner, 1990; Walsh, 1990).

Environmental Groups. Environmental groups were extremely supportive of all aspects of the materials separation provision and raised two issues in their comments on the proposal (Hershkowitz, 1990; Martineau, 1990b; Ruston, 1990). First, they suggested that the presence of metals can catalyze formation of dioxins/furans; however, no supporting data were provided. Second, they suggested that removing chlorinated plastics would reduce emission levels for HCl and dioxins/furans, again without including any supporting documentation. Additional benefits of materials separation cited by the Natural Resources Defense Council included (Doniger, 1990):

• Reduction in air pollution by burning less waste and removing items that may produce toxic emissions;
• Savings of millions of barrels of imported oil; and
• Reduction in future incineration ash landfill capacities.

Information with which to evaluate these claimed benefits, however, was not available.

MWC Industry. Representatives of the MWC industry echoed state and local government concerns about the lack of data demonstrating reduced emissions. They also questioned the legal authority of using the CAA to regulate materials separation and recycling. Industry believed that existing or new RCRA authority or a presidential recycling initiative would be more appropriate vehicles for increasing national recycling efforts (Institute of Resource Recovery, 1990; Martineau, 1990c).

Executive Branch. Within the executive branch, there also was a lack of agreement with EPA regarding anticipated benefits of the materials separation provision. The Council of Economic Advisers argued that EPA failed to establish evidence that separating materials would result in air pollution benefits. It is also noted that, as proposed, the materials separation requirement constituted a goal of the new source performance standards, not a tool that operators could use to meet a performance standard (Gruenspecht, 1990).

The President's Council of Competitiveness (PCC) joined the debate on this issue. In correspondence, the Council (PCC, 1990) affirmed its support for voluntary and market-based recycling programs. The Council expressed full support for stringent new standards for air emissions as required in the new source performance standards. However, it considered a materials separation provision to be inconsistent with the administration's regulatory principles. The Council's reasons for this position included:

• The materials separation requirement was not performance based, because it did not allow flexibility to select the most efficient ways for meeting the standard;
• A nationwide, uniform requirement violated the Federalism Executive Order (E.O. 12612), which required agencies to avoid federal regulations in areas reserved for state and local governments; and
• Benefits of this requirement did not exceed costs of implementing the program, thus failing to meet the criterion of E.O. 12291.

In response to these public comments, EPA stated in the preamble to the final rule that achievability was not an issue in its decision to remove the materials separation requirement. Rather, because of uncertainty over the net benefits of materials separation (i.e., reductions in pollutant emissions) and the potential economic impacts of the requirement, the decision was made not to include the provision in the final NSPS rule (EPA, 1991a). The agency clarified its opinion that emission benefits would result from reducing the amount of waste combusted, but acknowledged that these benefits could not be measured (EPA, 1991b).

OBSERVATIONS ABOUT THE USE OF INFORMATION

As indicated earlier, a regulatory agency has two stages at which it can apply current scientific and technological information in the regulatory process. The first stage is deciding whether a federal regulatory program is necessary. The second stage is development of proposed and final rules. The information available to an agency can be obtained from published literature, commissioned studies, or supplied by the regulated community.

In the case of MWC, EPA was not consistent in its use of information. In making a decision that federal new source performance standards were needed for municipal waste combustion, the agency actually ignored existing information about the potential risks associated with emissions from these facilities and about the sophistication of the technology being required under the PSD program. Unfortunately, the agency also ignored scientific data when it proposed a materials separation provision. However, EPA did use a range of technical and scientific information in proposing standards for air pollution controls. Although scientific, technical, and economic information relevant to all aspects of the regulatory process was available to the agency staff, there is clear evidence that political forces influenced decisions about the scope of the regulatory program.

Findings of this case study indicate that:

• Scientific, economic, and technical data related to regulatory decisions are necessary to balance political considerations; to counter political influences effectively the agency must have comprehensive and sound data at its disposal.
• Perspectives of governments about the use of these types of information vary among federal, state, and local officials and depend on the strength of political influence.
• Incentives for identifying new information also vary among the stakeholders in a regulatory process.

We would emphasize, however, that these findings presuppose that the agency is prepared to take a leadership role in opposing the political forces internal and external to it when available scientific, technical, and economic information indicates no justification for a specific political position. Unfortunately, there is little evidence to date that EPA can, or is willing, to be a strong leader against political tides.

Science versus Politics

In reaching an initial decision to proceed with a federal regulatory program for municipal waste combustion (i.e., giving this activity priority over development of regulations for other sources), EPA ignored available scientific and technical information. For example:

• Available information clearly indicated that permits issued by states and local regulatory agencies for construction and operation of new MWC facilities required installation of the best demonstrated technology and operating practices.
• These requirements had received limited opposition from industry and considerable support from host communities.
• Despite local concerns, there was no evidence to conclude that new facilities with state-of-the-art pollution control technology actually posed unacceptable risks to host communities or the surrounding environment. The EPA's own assessments indicated that risks associated with operation of these facilities were acceptable.
• No evidence existed to suggest that emissions from *new* MWC facilities would constitute a major national public health threat.

The EPA relied only on a contracted study that was narrowly focused on MWC practices. The Agency made no attempt to generate or compile more comprehensive data, for example, current relationships of recycling programs with MWC and comparisons of MWC emissions and risks within a national environmental context. In some instances, EPA ignored existing information, such as results of agency risk assessments.

Throughout the history of political actions related to MWC, EPA has lacked control of its agenda. In developing this rule, the agency was at the "mercy" of political forces from within and from outside. The use of any scientific or technical data took second place to these political forces.

However, the political factors surrounding MWC were conflicting in nature. Concern about management of solid waste began to gain momentum at this time. Landfill space was becoming scarce. Municipal wastes were being transported across state boundaries for disposal. The desire of communities to use incineration was increasing. In contrast, there also was growing dissatisfaction with the lack of pollution controls on *existing* municipal waste facilities. Supporters of recycling and waste reduction programs were becoming increasingly concerned that construction of new MWC capacity would have a detrimental impact on new recycling and reduction efforts. In this political climate, the agency did not have the internal commitment to make decisions based on scientific and technical information.

The second point at which EPA ignored science and technology was in its decision to propose a materials separation requirement as part of a facility permit. At the time of development of the proposed rule, no data existed to support a claim that separation of materials would reduce toxic emissions. Agency staff have suggested that newly appointed EPA decision makers wanted to use a major rule to illustrate its commitment to recycling and reduction programs for solid waste management. Technical staff at EPA had no means to balance this internal political goal.

While politics shaped this part of the proposed standards, politics also led to the elimination of the materials separation provision in the final rule. Only through the intervention of offices in the executive branch was the internal agency conflict resolved. This political influence, combined eventually with an acknowledgement by EPA of the insufficiency of the data, resulted in the withdrawal of the materials separation requirements. Without strong political backing, however, scientific and technical information about the inadvisability of the provision would have continued to be ignored.

Different Governmental Perspectives

Our findings suggest that the type of technical and scientific information used may be different at each level of government. However, it is clear that decisions about the need for a regulatory program are dictated by political forces regardless of the jurisdiction. Once a need has been established, the range of stringency in regulatory requirements may differ. For example, local regulatory agencies have the most to gain by seeking and requiring stringent air pollution controls and the highest standards for operation of facilities, regardless of actual risks associated with these facilities.

Local agencies and political leaders are responsible for meeting the waste management needs of their community and must have the community's support in any action aimed at addressing these needs. Without applying the best available technology, communities in which these facilities are to be sited do not give this support. In addition, local site conditions and specific community demands are driving forces behind implementation of most standards for pollution control at MWC facilities. As a result, local environmental agencies have a major incentive to identify new data and information about risks, demonstrated technology, and implementation costs. This information enhances the credibility of the local agency's choice for regulatory requirements.

At the state level, it appears that there may be less urgency than that perceived at the local level. Waste management is not the responsibility of state agencies; however, protection of public health, broadly defined, is. State agencies must evaluate all conditions within their jurisdiction to determine feasible and cost-effective protection throughout their state. Therefore, the need to specify a level playing field through state requirements for MWC must be balanced with a need for communities to be able to make potentially difficult waste management choices.

The federal perspective must be even more broadly defined than that of the state. The EPA must develop standards that can be implemented at a national level but do not constrain local options for waste management. If national regulations are overly stringent, some communities could be precluded from using MWC as a viable alternative to other management options. However, the federal agency must also balance this need for state and local flexibility with a need to maximize protection of public health and the environment, both within and among states.

Thus, there are different perspectives among federal, state, and local governments when it comes to promulgating regulations. State and local officials can identify the most appropriate and often most stringent requirements on a case-by-case basis. Differences among facilities and locations might be justified. A federal regulatory program, however, must promulgate national standards; therefore, case-by-case analysis is not appropriate. For most regulatory programs, rules based on best demonstrated technology at the federal level constitute the "best that can consistently and easily be achieved." There is necessarily a need to "average" these requirements to provide maximum national protection at reasonable costs. If state and local environmental programs are given responsibility (with federal guidance), then facility requirements can be as stringent as local conditions make necessary. Given the extent of public scrutiny of MWC activities, it is unlikely that minimal health and environmental protection will be imposed in host communities.

Incentives for Seeking New Information

Incentives for seeking new information vary throughout the regulatory process and among the various stakeholders. The basis of the Clean Air Act, that is, implementation of technological and operational standards, should provide maximum incentives for EPA to seek out the best available information. To a limited extent, the agency did seek new information during the initial stages of evaluating MWC by commissioning a large multifaceted study. However, once political factors begin to intrude, new information either was not sought or was ignored. For example, the risk evaluations performed in 1987 and 1989 suggested that MWC did not pose major risks at either a national or local level. The EPA did not use this information as a means of balancing political pressure and, thus, had little influence on the decision about the need for a regulatory program.

A major incentive for regulators in seeking comprehensive and new information is to enhance their credibility and the enforceability of a regulatory program. The materials separation requirement in the proposed rule is an excellent example of the influence of limited and less credible data. Acceptance by state and local governments and industry was not forthcoming because inclusion of this provision was not based on sufficient supporting data. If the agency had used scientific and technical information in reaching its decision about the advisability of the provision, a major controversy in the development of the NSPS rule might have been avoided. Unfortunately, political pressures forced it to ignore such information. Should the agency desire to revisit a materials separation requirement in other rules or in future revisions to this rule, it will be necessary to seek appropriate data to support assumptions about the reduction in air emissions associated with materials separation and the impact on public health and the environment.

The CAA itself inhibits to some extent the incentive to identify new data about new and cutting-edge technological innovations. The act specifically requires a standard of performance that is defined as the

> best system of emission reduction which (taking into account the cost of achieving such reduction . . .) the Administrator determines has been adequately demonstrated [CAA 42 USC 7411, Section 111(a)].

If a technology has not been used sufficiently to demonstrate consistent emission reductions, it is unlikely to be considered "demonstrated" because there is no opportunity for the agency to search beyond systems currently in operation. Statutes such as the CAA result in the development of *technologically driven* regulations. By specifying consistent achievements as the definition of performance, EPA can only look at those technologies used at existing facilities. The agency cannot promulgate *technology-forcing* regulations—those that force development of new technologies capable of achieving emissions standards beyond what is possible with current technology—under the CAA. To some extent the agency does attempt to force new technologies by setting standards at the upper bound of margins of achievability. However, the definition of BDT limits this option.

Industry also has limited incentives to develop new technology. In conversations with industry representatives, it was emphasized that industrial incentives for technological improvements in air pollution or operational control systems are to reduce operation and compliance costs.[6] Reduced emissions may be an added benefit of any cost reduction, but they are not the primary goal of new technical research. If a new technology can reduce emissions but is more costly to operate, industry would be less likely to implement it voluntarily. Until all facility operators are required through rulemaking to install the same, more costly equipment, such voluntary actions place operators with more advanced thinking at an economic disadvantage. The main reason that there was so little opposition to most of the provisions in the proposed new source performance standards is that most new facilities were already being required to implement the equipment and practices recognized as BDT in the proposed rule. Thus, the rule simply equalized the playing field within the regulated community.

Environmental groups probably have the greatest incentive to seek new information. In general, the agenda of these groups is to drive regulations to ever more stringent levels in a desire to maximize protection of public health and the environment. These groups often are not constrained by concerns for technical feasibility or economic factors. Therefore, they can identify new developments before a technology is considered to be technically or economically viable. Unfortunately, these groups may become so focused on a perceived need for more stringent requirements that they may not support their allegations about technical feasibility or risks posed by unregulated facilities with sound scientific and technical information.

CONCLUSIONS

The environmental regulatory agenda has always been shaped by political forces. Since the inception of EPA in 1970, Congress, industry, and public interest groups have identified various industrial activities thought to require a regulatory program. For example, in the early 1970s, Congress seemed to identify a different pollutant every year, which, in its opinion (or the opinion of a few members), would lead to dire public health and environmental problems without the immediate and rigorous attention of EPA. Industry and environmental groups often were sources of information that supported such urgency. Environmental and other public interest groups also have used specific environmental issues to further membership drives or to force a change in national policy, for example, the recent Alar controversy. Likewise, industry has lobbied Congress and EPA for regulatory programs that would provide more uniform regulations across the country, thus preventing uneven economic advantages or enhancing the ability to implement better technology.

In all such instances, regardless of the origination of an initiative, we are left with the impression that scientific and technical information rarely plays a determining role in the debates. Unfortunately, all stakeholders attempt to limit full use of scientific and technical information in the regulatory decision-making process, particularly if the information is counter to the stakeholder's agenda.

The outcome of this political struggle is an increased likelihood that significant problems, and important scientific data, are ignored. This point has been emphasized by the Expert Panel on the Role of Science at EPA (EPA, 1992). The panel stated in its report:

> Science is also key to determining which environmental problems pose the greatest risks to human health, ecosystems, and the economy. In the absence of sound scientific information, it is likely that high-profile but low risk problems will be targeted, while more significant threats are ignored. . . . Strong science provides the foundation for credible environmental decisionmaking (EPA, 1992, p. 15–25).

There is no question that politics both shaped the decision about whether a federal regulatory program was necessary for MWC and influenced provisions in the proposed and final new source performance standards. Whether, and how, such political forces can be curbed is questionable and may not be altogether desirable. However, based on the findings of this case study, certain suggestions emerge that may facilitate a better balance between political considerations and use of scientific, technical, and economic information in the regulatory decision-making process.

First, politics should be balanced with scientific, economic and technical information. In this case study, EPA evaluated such information related to MWC and published its findings in the 1987 report to Congress. However, the agency evaluated MWC as a single issue divorced from the broader national environmental context. Thus, while the agency estimated emissions from existing and projected facilities, it is not apparent that these emissions were compared with other industrial sources to determine their significance from a national perspective. Similarly, when the risks from exposure to these emissions were calculated, they were not compared with risks associated with other sources of these same chemicals and metals.

The EPA's economic analysis also failed to put the MWC issue into a national context. The agency considered costs for controlling emissions only. There was no attempt to compare costs of the federal program in broad terms with *national* environmental and public health benefits. The EPA should conduct a cost-benefit analysis using several competing environmental issues to determine if a specific problem (i.e., MWC) merits a federal program or whether there are alternative and less costly regulatory options.

Analysis of the relative national priority, evaluating consequences of not having a federal program, the cost of a federal program, and other problems competing for limited resources may result in a more effective balance of political pressure. A stronger foundation for balancing political influences might have been achieved if EPA had conducted such analyses when reaching a decision about the need for a federal MWC regulatory program. As it turned out, EPA staff had little real information with which to counter effectively the political pressures being brought to the MWC debate.

A second recommendation concerns a mechanism to enhance the quality of information available to the agency for use in developing regulatory requirements. Both government and industry agree that, for the most part, EPA used available information in formulating the proposed and final rules for new source performance standards. As indicated in an earlier discussion, the standards are largely based on the technological capability of the best demonstrated technology. However, conversations with EPA officials indicate that the information-gathering process might be enhanced (personal communication with R. Brenner and J. Democker, EPA Office of Air and Radiation, 1992). At present, EPA believes that stakeholders are playing a passive role in the agency's search for necessary information. Only in limited instances do stakeholders voluntarily provide information before a rule is proposed. Much information often appears to be withheld until the public comment period for a proposed rule.

Furthermore, state and local governments, environmental groups, and industry often complain that information used by the agency is not always of the highest quality. This complaint is supported by an observation of the Expert Panel on the Role of Science in EPA:

EPA program offices often conduct scoping studies or other preliminary assessments in the early stages of regulatory development. These studies are frequently carried out without benefit of peer review or quality assurance. They sometimes escalate into regulatory proposals with no further science input, leaving EPA initiatives on shaky scientific ground and affecting the credibility of the Agency (EPA, 1992; p. 37).

The EPA did conduct a large-scale study of MWC. However, to our knowledge, with the exception of an EPA Science Advisory Board review of only the methodology used in the risk assessment work, it was not peer reviewed (Hartung and Nelson, 1987). During development of a proposed standard, the only opportunity for any interested party to provide EPA with better information is to provide data informally and voluntarily. The extent and quality of information provided in this manner depends on the awareness of different stakeholders about directions the agency may take, or is taking, in developing a proposed rule. In general, EPA provides limited information about policy choices while a proposed rule is being developed, thus stakeholders are not always aware of gaps in the agency data base. Once a rule is published for public comment, additional information can be forthcoming, and EPA can apply it in revisions for the final rule. This seems to be an inefficient process that fosters adversarial positions rather than constructive comments.

A more effective mechanism is needed through which the agency can request and receive new information during *development of a proposed rule*, rather than waiting to receive information submitted voluntarily or during the public comment period. The Expert Panel on the Role of Science has provided extensive suggestions for enhancing the use of better information (EPA, 1992).

An additional recommendation is to employ regulatory negotiations or some version of these negotiations to enhance acquisition of better information. Regulatory negotiations have been used by EPA over the past five years in selected rulemaking to reduce adversarial reactions to proposed rules. The concept is to bring together representatives of all major interest groups for a specific rule and to develop a proposal to which all representatives can agree. Some of these negotiating attempts have been successful, resulting in proposed rules that enjoyed the consensus of all stakeholders. However, even when an acceptable proposal is not negotiated, there is still some success in terms of the amount of scientific and technical information that is brought into the negotiation process early.

To illustrate, a regulatory negotiation was attempted in developing RCRA regulations for deep-well injection of hazardous wastes.[7] Because of philosophical differences among the various interests at the negotiation table, consensus on a proposed rule was not achieved. However, most participants in this particular process acknowledged that a wealth of information was brought to the agency's attention. In the absence of these regulatory negotiations, it is questionable whether the full extent of pertinent information would have been presented through the normal voluntary mechanisms.

One reason that there is an increase in the volume of new information provided during such negotiations is that all participants are able to learn firsthand the basis for any opposing views. Thus, new information relevant to specific issues can be brought to the negotiating table as a counter to such objections. If negotiations were to become a standard component of a regulatory process, stakeholders would have a more controlled and focused opportunity to provide EPA with information in an atmosphere of cooperation. In addition, the quality and direct relevance of the scientific and technical information might be substantially enhanced. The ultimate result would be a stronger foundation for the development of scientifically and technically sound regulations.

As our findings suggest, EPA can benefit from implementing a process that enhances not only the quality of data collected but also the range of information. If informed regulatory decisions are to be made, current scientific and technical information is needed to balance political influences. At the first stage of the regulatory process, this information must be evaluated in a national context. At the second stage of regulatory development, the scope of information must be sufficiently broad to allow development of sound, achievable, and enforceable regulatory standards.

While these recommended options offer other mechanisms for enhancing the availability of new scientific and technical information to the agency, they do not address the underlying and extremely critical problem inherent in a regulatory process: the influence of political factions to the exclusion of scientific and technical information. The MWC rule for new source performance standards is not unique in the role that politics played in determining the need for and the scope of the regulatory program. Politics have "interfered" in many of the agency's activities. Such political pressure arises externally and internally to EPA. Until political factors are placed on an equal, rather than superior, footing with scientific, economic, and technical factors, EPA—and all regulatory agencies—will not be able to function effectively. Unfortunately, the resolution of this problem is not simple and will require a commitment on the part of all stakeholders to allow such information to play a more prominent role.

APPENDIX:
DESCRIPTIONS OF AIR POLLUTION CONTROLS

Electrostatic Precipitators

Electrostatic precipitators (ESPs) are used to remove particulate matter from flue gas streams. A typical ESP consists of an alternating array of negatively charged grids of wires and positively charged collection plates. Incoming particles are given an electrical charge through contact with gas ions. Charged particles pass through a strong electronic field, which causes these particles to migrate to a collection electrode with an opposite charge.

The precipitators are divided into sections called fields; adding fields increases the collection efficiency (Frillici and Schwartz, 1991).

Fabric Filters

Fabric filters (FFs) mechanically separate particles from flue gas streams, achieving greater than 99 percent removal (Gaige and Halil, 1992). They consist of a filter medium, (i.e., tubular bag), a cage to support the bags, a gas-tight enclosure, and a mechanism to remove accumulated particles periodically. As the particulate-laden gas passes through the filter medium, collected material forms a porous cake, which acts as an additional filtration medium. Fabric filters are categorized according to how they are cleaned: shaker, reverse-air, and pulse-jet. The fabric of the filters may either be woven or felted and may consist of fiberglass, Teflon or Nomex, which will withstand entering flue gas temperatures of up to 300°F (Frillici and Schwarz, 1991).

Acid Gas Scrubbers

Acid gas scrubbers operate by bringing acid gases into contact with alkali reagents, forming a neutralized salt solid that can be removed by ESPs or FFs. Acid gas scrubbers are categorized as wet scrubbers, dry scrubbers, or wet-dry scrubbers. Wet scrubbers use lime, limestone, or an alkali reagent and produce a wet bottom catch; their design might include venturi, spray, baffle, or packed tower. Dry scrubbers inject dry sorbent into the flue duct, resulting in a dry catch (Frillici and Schwartz, 1991).

In wet-dry scrubbers, commonly referred to as spray dryer absorbers (SD), flue gas enters the reaction vessel, where it is dispersed and put into spiral motion. A water-based slurry of alkali reagent is sprayed into the flue gas stream; the water evaporates and the reagent reacts with SO_2 and HCl to form salts. Use of FFs in conjunction with an SD results in a cake where the reagent and gases can react further, increasing efficiency (Frillici and Schwartz, 1991; Gaige and Halil, 1992).

Although acid gas scrubbers are intended primarily to remove HCl and SO_2, they also remove some organic and heavy metal pollutants (Brna and Kilgore, 1991). Reduced flue gas temperatures associated with scrubbers cause many volatilized metals and organics to condense, thus increasing removal efficiencies in the ESP or FF (Gaige and Halil, 1992).

Nitrogen Oxide Controls

Two main add-on technologies are currently in use across the world: selective catalytic reduction (SCR) and selective noncatalytic reduction (SNCR).

SCR has been used on coal- and oil-fired power plants in Japan and Europe. Ammonia is injected into the flue gas stream and the mixture is then passed through a catalyst (molybdenum, vanadium, titanium) bed where NO_x are converted to nitrogen gas. SCR operates at temperatures ranging from 500 to 800°F. Fouling caused by high particulate loading limits the potential for application of this technology (Frillici and Schwartz, 1991).

SNCR involves postcombustion injection of ammonia or urea to contact the flue gas. SNCR is most effective between 1600 and 2000°F, so the injectors are located in the upper portion of the furnace. The gas phase reaction between the NO_x and the injected ammonia or urea results in the production of nitrogen gas and water. Potential disadvantages include the difficulty of maintaining the optimal flue gas temperature in the injection zone. Also, ammonia may react with acid gases to form ammonia salts, which may corrode and foul downstream equipment or exit the stack as a visible plume (Frillici and Schwartz, 1991).

NOTES

1. We interviewed staff of particular MWC companies and the Integrated Waste Services Association, the trade association for the industry. Discussions were held with representatives of the U.S. Conference of Mayors and the Association of State and Territorial Solid Waste Management Officials. Staff of the EPA Office of Air and Radiation and Office of Air Quality Planning and Standards were interviewed.

2. The first legislative action for Clean Air occurred in December 1963 (P.L. 88-206), and was amended eighteen times between 1963 and 1990.

3. An acceptable risk range of 10^{-6} to 10^{-4} was established by the Office of Solid Waste and Emergency Response for use in the Superfund Program during the mid-1980s.

4. The following health parameters were used in Morrison (1989) for comparison with MWC emissions. HCl: EPA reference dose (RfD) of 7 $\mu g/m^3$; Hg: National Emission Standards for Hazardous Air Pollutants guideline of 1 $\mu g/m^3$; and Pb: National Ambient Air Quality Standards of 1.5 $\mu g/m^3$ (quarterly average).

5. The conflict between Congress and EPA over the initial proposal to develop health-based treatment standards for the 1984 HSWA provisions for land disposal restrictions undoubtedly set a precedent for EPA's decision about a basis for requesting MWC emissions.

6. Personal conversations with representatives of Integrated Waste Services Association, Ogden Martin Systems, and ABB Resource Recovery Systems.

7. Dr. Pirages was a participant in this regulatory negotiation.

REFERENCES

Abrams, R., H. G. Williams, A. Violet, and J. I. Lieberman. 1986. Petition for a determination regarding health effects for the establishment of emission regulations under §112 of the United States Clean Air Act. States of New York, Rhode Island, and Connecticut.

Brna, T. G., and J. D. Kilgore. 1991. The impact of particulate emissions control on the control of other MWC air emissions. Journal of the Air Waste Management Association 40(9):1324–1330.

Cheremisinoff, P. 1987. Resource recovery: A special report. Pollution Engineering, November:52–59.

Curling, D. S. 1990. Comments on the proposed NSPS for MWCs from the Southeastern Public Service Authority of Virginia, Chesapeake, Va., dated January 29, 1990. EPA Air Docket No. A-89-08, IV-D-77.

DePaul, F. T., and J. W. Crowder. 1988. Control of emissions from municipal solid waste incinerators. Prepared for Illinois Department of Energy and Natural Resources, Energy and Environmental Affairs Division, Springfield, Ill. ILENR RE-AQ-88/14.

Doniger, D. 1990. Letter from David Doniger, Senior Attorney, NRDC to George Bush, President, dated December 27, 1990.

Frillici, P. W., and S. C. Schwartz. 1991. BACT, MACT, and the act: What's going on? Waste Age 22(11):65–72.

Gaige, C. D., and R. T. Halil, Jr. 1992. Clearing the air about municipal waste combustors. Solid Waste & Power, January-February:12–17.

Goldstein, E. A., D. D. Doniger, A. K. Ahmed and M. D. Uva. 1986. Petition to the United States Environmental Protection Agency for the Regulation of Emissions from Municipal Solid Waste Incinerators. Washington, D.C.: Natural Resources Defense Council.

Greim, H. 1990. Toxicological evaluation of emissions from modern municipal waste incinerators. Chemosphere 20(3/4):317–331.

Gruenspecht, H. 1990. Memo from Howard Gruenspecht, Council of Economic Advisers, to James McRae, Office of Management and Budget, dated December 7, 1990.

Hartung, R., and N. Nelson. 1987. Letter to Mr. L. M. Thomas, U.S. EPA Administrator. Science Advisory Board, Washington, D.C.

Hershkowitz, A. 1990. Letter from A. Hershkowitz, Natural Resources Defense Council, to William K. Reilly, U.S. EPA Administrator, dated December 20, 1990.

Institute of Resource Recovery. 1990. Written statement of the Institute of Resource Recovery regarding the U.S. EPA's proposed rules fur municipal waste combustors, dated March 1, 1990.

Integrated Waste Services Association. 1992a. Waste-to-energy. Washington, D.C.

Integrated Waste Services Association (IWSA). 1992b. Survey of recycling and waste-to-energy activities. Washington, D.C.

Kiser, J. V. L. 1991. Municipal waste combustion in the United States: An overview. Waste Age 22(11):27–30.

Kiser, J. V. L., and D. B. Sussman. 1991. Municipal waste combustion & mercury: The real story. Waste Age 22(11):41–44.

Kiser, J. V. L. 1992. Municipal waste combustion in North America: 1992 update. Waste Age 23(11):26–36.

Kiser, J. V. L., and B. K. Burton. 1992. Energy from municipal waste: Picking up where recycling leaves off. Waste Age 23(11):39–46.

Martineau, R. J., Jr. 1990a. Memo from Martineau, Attorney, U.S. EPA Air and Radiation Division re: September 12, 1990 meeting of EPA Officials and National Association of Counties.

Martineau, R. J., Jr. 1990b. Memo from Martineau, Attorney, U.S. EPA Air and Radiation Division re: September 10, 1990 meeting between EPA officials and representatives of Natural Resources Defense Council.

Martineau, R. J., Jr. 1990c. Memo from Martineau, Attorney, U.S. EPA Air and Radiation Division re: November 6, 1990 meeting with representatives of Waste Management Inc. and EPA representatives on proposed MWC rule.

Morrison, R. M. 1989. Baseline risk analysis to support municipal waste combustor new source performance standard and emission guideline development. Memorandum to file. U.S. EPA Office of Air Quality Planning and Standards, Research Triangle Park, N.C. EPA Air Docket #A-89-08.

National Association of Counties. 1990. Comments of the National Association of Counties regarding the Environmental Protection Agency's New Source Performance Standards for Municipal Waste Combustors, Washington, D.C., dated February 28, 1990. EPA Air Docket No. A-89-08, IV-D-116.

National Solid Waste Management Association. 1991. Resource recovery in North America. Washington, D.C.

Nosenchuck, N. H. 1992. Personal communication. Director, Division of Solid Waste, New York Department of Environmental Conservation, Albany, N.Y.

Nosenchuck, N. H., and D. Bruckner. 1990. Comments on the proposed NSPS for MWCs from the Association of State and Territorial Solid Waste Management Officials, dated March 1, 1990. EPA Air Docket No. A-89-08, IV-D-71.

Pederson, W. F., Jr. 1987. Air pollution control. Chapter 6 in Environmental Law Handbook, 9th ed., Arbuckle et al., eds. Rockville, Md.: Government Institute, Inc.

Porter, J. W. 1990. Municipal Solid Waste Recycling: The Big Picture, speech before the U.S. Conference of Mayors Recycling Conference, March 29. (Reported in IWSA, 1992a)

President's Council on Competitiveness. 1990. Fact Sheet re: Recycling Requirement in the Municipal Waste Combustor Rule, dated December 19, 1990.

Reisch, M. S. 1992. SO$_2$ emissions trading rights: A model for other pollutants. Chemical and Engineering News 70(27):21–22.

Roffman, A., and H. K. Roffman. 1991. Air emissions from municipal waste combustion and their environmental effects. The Science of Total Environment 104:87–96.

Ruston, J. F. 1990. Comments of the Environmental Defense Fund on the U.S. Environmental Protection Agency's Proposed Standards of Performance for New Stationary Sources; Municipal Waste Combustors, Washington, D.C., dated March 1, 1990. EPA Air Docket No. A-89-08, IV-D-173.

Thomas, L. M. 1985. Statement before the Subcommittee on Health and the Environment and Commerce, U.S. House of Representatives. Washington, D.C.

U.S. Court of Appeals for the D.C. Circuit. 1992. No. 91-1168, State of New York and State of Florida v. W. K. Reilly, Administrator, U.S. Environmental Protection Agency, and No. 9101170, Natural Resources Defense Council v. W. K. Reilly, Administrator, U.S. Environmental Protection Agency.

U.S. Environmental Protection Agency (EPA). 1985. A strategy to reduce risks to public health from air toxics. Research Triangle Park, N.C.: Office of Air Quality Planning and Standards.

U.S. Environmental Protection Agency. 1987a. Municipal Waste Combustion Study: Report to Congress. Washington, D.C.: Office of Solid Waste and Emergency Response, Office of Air and Radiation and Office of Research and Development. EPA/530-SW-87-021a.

U.S. Environmental Protection Agency. 1987b. Response to petition for rulemaking and advance notice of proposed rulemaking. Federal Register 52(129):25399–25409.

U.S. Environmental Protection Agency. 1987c. Operational Guidance on Control Technology for New and Modified Municipal Waste Combustors. Research Triangle Park, N.C.: Office of Air Quality Planning and Standards.

U.S. Environmental Protection Agency. 1987d. Municipal waste combustion study: Assessment of health risks associated with municipal waste combustion emissions. Washington, D.C.: Office of Solid Waste and Emergency Response, Office of Air and Radiation and Office of Research and Development. EPA/530-SW-87-021g.

U.S. Environmental Protection Agency. 1989a. The Solid Waste Dilemma: An Agenda for Action. Final report of the Municipal Solid Waste Task Force. Washington, D.C.: Office of Solid Waste. EPA/530-SW-89-019.

U.S. Environmental Protection Agency. 1989b. Standards of performance for new stationary sources; municipal waste combustors; proposed rule. Federal Register 54(243):52251–52304.

U.S. Environmental Protection Agency. 1989c. Interim Procedures for Estimating Risks Associated with Exposures to Mixtures of Chlorinated Dibenzo-p-Dioxins and -Dibenzofurans (CDDs and CDFs) and 1989 Update. Washington, D.C.: Risk Assessment Forum. EPA/625/3-89/016.

U.S. Environmental Protection Agency. 1990. Municipal Waste Combustion: Background Information for Materials Separation. Research Triangle Park, N.C.: Office of Air Quality Planning and Standards. EPA-450/3-90-021.

U.S. Environmental Protection Agency. 1991a. Standards of performance for new stationary sources; municipal waste combustors; final rule. Federal Register 56(28):5488–5527.

U.S. Environmental Protection Agency. 1991b. Municipal Waste Combustion: Background Information for Promulgated Standards and Guidelines—Summary of Public Comments and Responses. Research Triangle Park, N.C.: Office of Air Quality Planning and Standards. EPA/450/3-91-004.

U.S. Environmental Protection Agency. 1992. Safeguarding the Future: Credible Science, Credible Decisions. Expert Panel on the Role of Science at EPA. EPA/600/9-91/050.

Walsh, D. C. 1991. The nation's first resource recovery plant. Waste Age 22(11):62–64.

Walsh, M. W., Jr. 1990. Comments on the proposed NSPS for MWCs from the Maryland Department of the Environment, Baltimore, Md., dated April 2, 1990. EPA Air Docket No. A-89-08, IV-D-277.

Waste Age. 1992. The 1992 municipal waste combustion guide. Waste Age 23(11):99–117.

Keeping Pace with Science and Engineering. 1993.
Pp. 141–164. Washington, DC: National Academy Press.

Trihalomethanes and Other By-Products Formed by Chlorination of Drinking Water

Philip C. Singer

Chlorine has been used to disinfect drinking water in the United States and in most of the world since 1908. Its widespread use has been credited with the control of a number of waterborne diseases, most notably typhoid fever and cholera. However, with the discovery in 1974 of trihalomethanes (THMs) in chlorinated drinking water and, subsequently, other halogenated disinfection by-products with potential adverse health impacts, the practice of chlorination has been seriously questioned. Trihalomethanes in finished drinking water have been regulated in the United States since 1979 and the U.S. Environmental Protection Agency (EPA) is considering adopting more stringent regulations for THMs; it may also establish maximum contaminant levels or treatment techniques for several other disinfection by-products (DBPs).

This paper reviews the scientific findings associated with the formation of THMs and other disinfection by-products and discusses how these findings have affected strategies for controlling these by-products in drinking water, and the corresponding improvement in the protection of public health. There are a number of confounding issues associated with the management and regulatory strategies for controlling THMs and the other disinfection by-products; these confounding factors are included in the discussion.

CHRONOLOGY OF SCIENTIFIC FINDINGS

Figure 1 presents a chronology of the more noteworthy scientific findings involving the formation of THMs and other disinfection by-products

141

Science and Engineering   **Policy and Regulation**

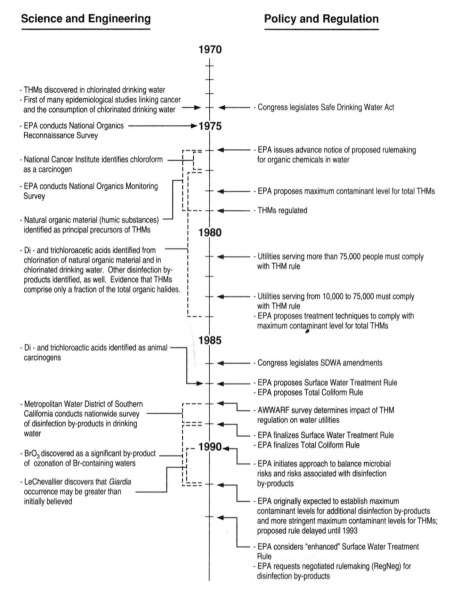

FIGURE 1   Timeline of significant scientific and regulatory events for trihalo-methanes and other chlorination by-products.

generated during the treatment of drinking water. Trihalomethanes were first identified in finished drinking water in 1974, both in the Netherlands in Rotterdam (Rook, 1974) and in the United States in New Orleans, Louisiana (Bellar et al., 1974). Their presence was linked to the practice of chlorinating water. In 1975 the U.S. Environmental Protection Agency conducted the National Organics Reconnaissance Survey of 80 cities in the United States and found that the four THMs—chloroform ($CHCl_3$), bromodichloromethane ($CHBrCl_2$), dibromochloromethane ($CHBr_2Cl$), and bromoform ($CHBr_3$)—occurred widely in chlorinated drinking water and resulted from the practice of chlorination (Symons et al., 1975). Figure 2 shows that in the 80-city survey, total trihalomethane concentrations in the finished drinking water correlated with the nonpurgeable organic carbon concentrations in the raw water. From March 1976 to January 1977 the EPA conducted the National Organics Monitoring Survey, which verified the findings of the earlier survey, and demonstrated that THMs continued to form to a significant extent in the finished water distribution system (Brass et al., 1977).

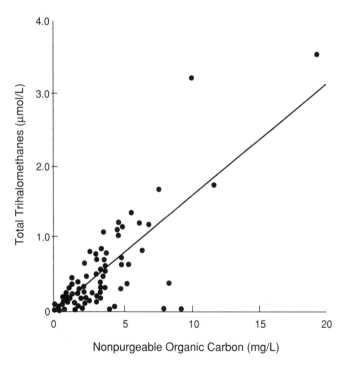

FIGURE 2 Relationship between trihalomethane formation in finished water and nonpurgeable organic carbon in source water. SOURCE: Symons et al. (1981).

A number of studies conducted in the late 1970s and early 1980s indicated that many other halogenated by-products also formed in drinking water as a result of chlorination, in addition to the THMs. These studies were conducted by chlorinating raw drinking water and humic material, the principal organic component of most natural waters, and by making measurements in finished drinking water. The most frequently identified disinfection by-products, in addition to the THMs, were di- and trichloroacetic acid, di- and trichloroacetonitrile, chlorinated ketones, chloral hydrate, and chloropicrin (e.g., Christman et al., 1983; Coleman et al., 1984; Miller and Uden, 1983; Oliver, 1983; Quimby et al., 1980; Reckhow and Singer, 1984: Rook, 1977; Trehy and Bieber, 1981). Numerous other halogenated disinfection by-products have been identified, but with less frequency and at trace levels (e.g., Stevens et al., 1989). The formation of halogenated disinfection by-products from the reaction between chlorine and natural organic material (NOM) is given by the following general equation:

$$Cl_2 + NOM \rightarrow CHCl_3 + \text{Other THMs} + \text{Other DBPs}$$

Despite the fact that researchers have identified hundreds of halogenated disinfection by-products in chlorinated water, the total concentration of those compounds that have been quantified amounts to only about 50 percent of the total organic halide (TOX) content (e.g., Christman et al., 1983; Reckhow and Singer, 1984; Singer and Chang, 1989). By separately measuring the total organic halide concentration in chlorinated water using an adsorption/pyrolysis/coulometric detection procedure (Standard Methods, 1985), researchers have demonstrated that the sum of the measured THMs, haloacetic acids (HAAs), haloacetonitriles, etc., when converted to chlorine-equivalent concentrations, accounts for only about 50 percent of the measured total organic halide concentration, also in chlorine-equivalent units. This means that approximately half of the halogenated disinfection by-products consist of unidentified halogenated compounds.

In 1988-89 the Metropolitan Water District of Southern California and James M. Montgomery Consulting Engineers, in a project jointly sponsored by the EPA, the California Department of Health Services, and the Association of Metropolitan Water Agencies, conducted the most comprehensive survey of disinfection by-products in finished drinking water to date (Krasner et al., 1989; McGuire et al., 1989). The investigators analyzed finished drinking water in 35 utilities nationwide, 10 of which were in California, for a variety of disinfection by-products for which analytical techniques were available (see Table 1). The study was directed at detecting disinfection by-products, seasonal patterns in their formation, the effects of raw water quality on the levels and distribution of the by-products analyzed, and effects of treatment modifications on by-product formation.

TABLE 1   Targeted Compounds in Nationwide
Disinfection By-Product Survey

Trihalomethanes
    Chloroform
    Bromodichloromethane
    Dibromochloromethane
    Bromoform

Haloacetic acids
    Chloroacetic acid
    Dichloroacetic acid
    Trichloroacetic acid
    Bromoacetic acid
    Dibromoacetic acid

Haloacetonitriles
    Dichloroacetonitrile
    Trichloroacetonitrile
    Bromochloroacetonitrile
    Dibromoacetonitrile

Haloketones
    1,1-dichloropropanone
    1,1,1-trichloropropanone

Miscellaneous chloro-organics
    Chloropicrin
    Chloral hydrate
    Cyanogen chloride
    2,4,6-trichlorophenol

Aldehydes
    Formaldehyde
    Acetaldehyde

SOURCE: McGuire et al. (1989).

The results of this survey indicated that THMs were the by-products present in the highest concentrations in finished drinking water, with the haloacetic acids present at approximately 50 percent of the total THM concentration. The mean and median total THM concentrations were 39 and 36 micrograms/liter (μg/L), respectively, while the median total haloacetic acid concentration was 17 μg/L. [Recent research by J.M. Montgomery Consulting Engineers (1992) and Singer et al. (1992) suggests a higher ratio of haloacetic acids to trihalomethanes.] A number of water samples had significant concentrations of by-products containing bromine because the source waters had high concentrations of bromide ion.

As a result of all of these findings, epidemiologists and toxicologists have conducted numerous studies in an attempt to evaluate the impact of chlorination on public health and, particularly, the health effects of the specific halogenated disinfection by-products that have been identified. A large number of epidemiological studies have been conducted in the United States since 1974. These studies have repeatedly shown a weak association between the chlorination of drinking water and an increased incidence of cancer, but a causal relationship between exposure to chlorinated drinking water and cancer has not be established (Craun, 1991). Most of the historical studies have demonstrated weak relationships between bladder, colon, and rectal cancers and the consumption of chlorinated drinking water, but more recent studies have also shown evidence of relationships between pancreatic cancer (IJsselmuiden et al., 1992) and birth defects (Bove et al., 1992a,b) with the consumption of chlorinated drinking water. Although there is no single study that can be cited as a seminal study linking chlorination and cancer, it is the sheer weight of evidence provided by the large number of studies showing a positive relationship, albeit a weak one, that underscores the concern, from a cancer risk perspective, about the safety of drinking chlorinated water (Morris et al., 1992). These epidemiological studies have been extensively reviewed by Bull and Kopfler (1991), Craun (1988, 1991), and the National Research Council (NRC, 1980, 1987).

From a toxicological viewpoint, chloroform has been shown to induce liver tumors in mice and kidney tumors in rats (Jorgenson et al., 1985; National Cancer Institute, 1976). Bromodichloromethane has been shown to induce renal tumors in mice and rats, liver tumors in mice, and intestinal tumors in rats (National Toxicology Program, 1986). Bromoform produced intestinal tumors in male and female rats (National Toxicology Program, 1989). Dichloroacetic acid and trichloroacetic acid induced the formation of hepatic tumors in mice (Bull et al., 1990; Herren-Freund et al., 1987). Accordingly, based on these animal studies, it can be concluded that a number of halogenated by-products formed during the chlorination of drinking water are probable human carcinogens. The National Research Council (1987) and Bull and Kopfler (1991) recently reviewed and profiled the health effects of a number of disinfectants and disinfection by-products. According to Bull and Kopfler (1991), the only halogenated by-products that appear to approach concentrations of regulatory concern in chlorinated drinking water are the four THMs, di- and trichloroacetic acid, chloropicrin, and trichlorophenol, although the EPA has generated a different list of candidate compounds for regulation, as illustrated in Table 2 (Regli et al., 1992).

Another class of halogenated by-products, the halogenated furanones, exemplified by MX [3-chloro-4-(dichloromethyl)-5-hydroxy-2(5H)-furanone], have been found in chlorinated drinking water (Kronberg et al., 1988) at concentrations on the order of 50 nanograms/liter (0.050 µg/L). Despite

TABLE 2   Candidate Disinfection By-Products for Regulation

Compound	Health Effects and Cancer Status	Possible Maximum Contaminant Level Goal
Trihalomethanes		
Chloroform	Cancer, B2	0
Bromodichloromethane	Cancer, B2	0
Dibromochloromethane	Liver, C	60 µg/L
Bromoform	Cancer, B2	0
Haloacetic Acids		
Trichloroacetic Acid	Liver, C	100 µg/L
Dichloroacetic Acid	Cancer, B2	0
Other		
Chloral Hydrate	Liver, C	5 µg/L
Bromate	Cancer, B2	0
Chlorine	Blood, D	4 mg/L
Chloramines	Blood, D	4 mg/L
Chlorine Dioxide	Blood, Neurological, D	0.8 mg/L
Chlorite	Blood, D	0.3 mg/L

CODE: B2 = Probable human carcinogen; C = Possible human carcinogen; D = Inadequate or no evidence of human or animal carcinogenicity

SOURCE: Regli et al. (1992).

their low concentrations, they have been shown to be responsible for up to 50 percent of the mutagenicity of chlorinated drinking water. The significance of this class of compounds to public health has been questioned, however, because of the likelihood that they are detoxified in mammalian systems following ingestion. The link between mutagenicity and human carcinogenicity is a subject of scientific debate.

In fact, interpretations of the health effects studies overall have been extensively criticized by scientists, engineers, and water utility managers. In 1991 the International Agency for Research on Cancer concluded that there was inadequate evidence for the carcinogenicity of chlorinated drinking water in humans or laboratory animals and that chlorinated drinking water was not classifiable as to its carcinogenicity (World Health Organization, 1991). The issues involve the weakness of the reported epidemiological relationships between cancer and consumption of chlorinated drinking water, the high dosages of test compounds administered to laboratory animals to induce tumors, and the validity of the models used to extrapolate from these high dosage effects to the low concentrations at which these compounds are found in drinking water. These are issues common to many environmental contaminants. There is little question that estimation of the public health risk involved in consuming chlorinated drinking water and in establishing maximum contaminant levels for disinfection by-products is fraught with uncertainty.

CHRONOLOGY OF REGULATORY ACTIONS

The chronology of regulatory actions taken in response to the scientific discoveries concerning the formation of trihalomethanes in chlorinated drinking water and the associated health concerns is also shown in Figure 1. Following passage of the Safe Drinking Water Act by Congress in 1974 and the findings of the National Organics Reconnaissance Survey, the Environmental Protection Agency published its advance notice of proposed rulemaking on July 14, 1976 to address control options for organic chemical contaminants in drinking water (EPA, 1976). Two approaches were considered: establishing maximum contaminant levels for specific organic chemicals or for surrogates (indicators) of these organic chemicals, and establishing designated treatment techniques to control specific organic contaminants or their surrogates.

On February 9, 1978, the EPA published a proposed rule to amend the National Interim Primary Drinking Water Regulations to include a maximum contaminant level and associated monitoring and reporting requirements for total trihalomethanes (EPA, 1978). At the same time, the EPA proposed a requirement for the use of granular activated carbon or equivalent technology for application to drinking water that was presumed to be vulnerable to contamination by synthetic organic chemicals of industrial origin.

Following a period of public comment, on November 29, 1979, the EPA adopted its final rule for the control of THMs in drinking water (EPA, 1979). The rule amended the National Interim Primary Drinking Water Regulations by establishing a maximum contaminant level for total THMs of 0.10 milligrams/liter (mg/L) (100 µg/L). For community water systems serving 75,000 or more persons, the effective date of compliance with the maximum contaminant level was November 29, 1981, and for community water systems serving 10,000 to 75,000 persons, the effective date of compliance with the maximum contaminant level was November 29, 1983. Compliance for systems serving fewer than 10,000 customers was left to the discretion of the individual states. The THM rule also established monitoring and reporting requirements that revolved around the collection and analysis of samples from representative locations in the water distribution system on a quarterly basis. The rule stipulated that the running annual average of the arithmetic sum of the concentrations of all four THM species had to be less than or equal to 0.10 mg/L.

The earlier proposed requirement for the use of granular activated carbon for vulnerable water supplies was not included in the final THM rule, but was subsequently adopted as part of EPA's regulations controlling synthetic organic chemicals in drinking water. [The Safe Drinking Water Act Amendments of 1986 declared granular activated carbon treatment to be the best available technology (BAT) for the control of synthetic organic chemicals.]

The EPA's selection of an interim maximum contaminant level of 0.10 mg/L was based on balancing public health considerations against the technological and economic feasibility of limiting total THM concentrations to such levels in public water systems in the United States. In addition, the limited data base available at that time, from the standpoint of both occurrence and health effects, prevented the EPA from establishing individual maximum contaminant levels for each of the four THM species. Finally, the EPA did not extend the maximum contaminant level to community water systems serving fewer than 10,000 persons because of concerns about the technical and economic feasibility of such systems being able to comply with the rule without jeopardizing their disinfection practices and putting their customers at an increased risk of waterborne diseases.

It was assumed that many water utilities would ultimately be able to achieve total THM concentrations as low as 0.010 to 0.025 mg/L (10-25 μg/L), and EPA suggested these values as future goals to be considered in the Revised National Primary Drinking Water Regulations.

On February 28, 1983, in accordance with the stipulations of the Safe Drinking Water Act, the EPA identified the best technology and treatment techniques that community water systems could use to achieve compliance with the maximum contaminant level for total THMs (EPA, 1983). These techniques were believed to be "generally available, taking costs into consideration." The specific techniques identified were the use of chloramines or chlorine dioxide as alternative or supplemental disinfectants and oxidants, improved clarification for THM precursor removal, moving the point of chlorination to reduce THM formation, and the use of powdered activated carbon to remove THMs or THM precursors. Additional techniques not determined to be "generally available" included the use of ozone as an alternative or supplemental disinfectant or oxidant, aeration for THM removal, off-line water storage, consideration of alternative sources of raw water, and implementation of clarification if not currently practiced.

Following promulgation of the THM rule and the setting of a maximum contaminant level for total THMs in finished drinking water, the water treatment industry responded by adopting practices to limit THM formation. The principal treatment modifications involved moving the point of chlorination downstream in the water treatment plant, optimizing the coagulation process to enhance the removal of THM precursors, and using chloramines to supplement or replace the use of free chlorine. Young and Singer (1979) had shown that moving the point of chlorine application from the raw water to a location after clarification could reduce THM formation by approximately 40 percent. A number of other researchers (e.g., Babcock and Singer, 1979; Johnson and Randtke, 1983; Kavanaugh, 1978; Semmens and Field, 1980) had demonstrated that up to 75 percent of the THM precursors could be removed by coagulation, sedimentation, and filtration if the coagulant dose and pH were optimized. Other researchers (e.g., Brodtmann et al., 1980; Duke et al., 1980; Lange and Kawczynski 1978; Norman et al., 1980) had shown that the addition of ammonia to water containing chlorine essentially stopped the subsequent formation of THMs. This made the use of chloramines a very simple, inexpensive, and therefore attractive approach for limiting THM formation. However, because chloramines are a much weaker disinfectant than free chlorine, concern about compromising the microbiological quality of drinking water became an important issue.

McGuire and Meadow (1988) surveyed 727 water utilities for the American Water Works Association Research Foundation (AWWARF) to determine the extent to which utilities were in compliance with the maximum contaminant level for total THMs, and the cost of achieving such compliance. The results showed that enactment of the THM regulation resulted in a 40-50 percent average reduction in total THM concentrations for the larger utilities surveyed. The median total THM concentration among all the respondents was 38 µg/L, a value not much different than the results of EPA's National Organics Reconnaissance Survey (Symons et al., 1975) or the National Organics Monitoring Survey (Brass et al., 1977). The principal differences were that the utilities with high THM levels were able to reduce their THM concentrations substantially, as shown in Figure 3. Of those systems that implemented THM control measures, the majority did one or more of the following things: (1) modified their point(s) of chlorine application, (2) changed their chlorine dosages, and (3) adopted the use of chloramines to ensure compliance with the maximum contaminant level for total THMs. A large number of utilities changed from free chlorine to chloramines as their primary disinfectant. Table 3 summarizes the treatment modifications made by the utilities included in the survey. Although compliance with the 0.10 mg/L (100 µg/L) maximum contaminant level was found to be not particularly costly, it was concluded that reducing the maximum contaminant level significantly below 50 µg/L would cause a large number of utilities to exceed the maximum contaminant level and would require extensive capital expenditures to bring these utilities into compliance with such a more stringent value.

In summary, the scientific findings on THMs and other halogenated disinfection by-products in chlorinated drinking water led to health effects studies that showed that chloroform and other disinfection by-products were

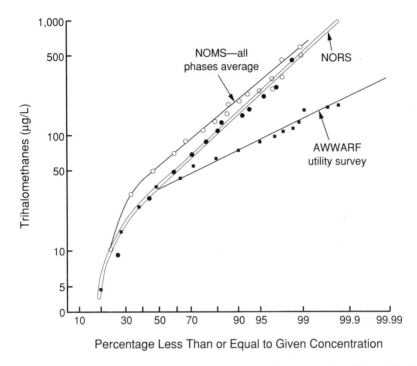

FIGURE 3 Frequency distribution of survey data from American Water Works Association Research Foundation (AWWARF) Utility Survey compared with the National Organics Reconnaissance Survey (NORS) and the National Organics Monitoring Survey (NOMS). SOURCE: McGuire and Meadow (1988). Reprinted from the Journal of the American Water Works Association by permission. Copyright © 1988, American Water Works Association.

carcinogenic in laboratory animals and that people drinking chlorinated water might be at somewhat greater risk in developing bladder, colon, and rectal cancer than those consuming unchlorinated drinking water. Additional scientific findings involving the formation and behavior of THMs and the other disinfection by-products led to the development of water treatment strategies to limit the formation of halogenated disinfection by-products. The enactment of the THM regulation resulted in treatment modifications that subsequently reduced the extent of THM formation and, most likely, of many of the other disinfection by-products as well, thereby lowering the public health risk associated with THMs and the consumption of chlorinated drinking water. Implementation of some of these modifications, however, may have compromised the microbiological quality of drinking water (see later discussion).

TABLE 3   Treatment Changes Made by Utilities to
Comply with 0.10 mg/L Maximum Contaminant Level
for Total Trihalomethanes

Treatment Change	Number of Plants
Change point of Cl_2	150
Change Cl_2 dose	122
Add NH_2Cl	100
Improve clarification	41
Add ClO_2	26
Add powered activated carbon	25
Install granular activated carbon	10
Install aeration	9
Install clarification	4
Provide off-line storage	4
Add O_3	3
Use alternate source	2
Other	47
Total	543

SOURCE: McGuire and Meadow (1988).

CHRONOLOGY OF CONFOUNDING EVENTS

A number of additional regulatory actions and scientific discoveries have taken place within the past six years that have served to confound the disinfection by-product issue. They revolve primarily around the 1986 amendments to the Safe Drinking Water Act, the promulgation by the EPA in 1989 of the Surface Water Treatment Rule and the Total Coliform Rule, the perceived need to establish maximum contaminant levels for other disinfection by-products in addition to the THMs, and the previously stated need (EPA, 1979) to reevaluate the maximum contaminant level for total THMs based on experiences gained following the initial regulation. It had been understood that the initial maximum contaminant level for total THMs was a compromise between public health considerations and technological and economic feasibility. These additional considerations, some of which are scientific and some of which are regulatory in nature are shown in Figure 1. The discussion that follows presents these events in the order in which they occurred, their relationship to the disinfection by-product issue, and the resultant complexity of management and control options for disinfection by-products.

The 1986 amendments reflected the dissatisfaction of Congress with the slow pace and manner by which the original Safe Drinking Water Act of 1974 was being implemented, growing public concern over contamination of drinking water supplies, and public perception that drinking water was unsafe. The key elements of the amendments with respect to the disinfection by-product issue were the development of specific requirements for (1) the disinfection of all public water supplies, including mandatory filtration of all surface water supplies; (2) the adoption of maximum contaminant levels for 83 contaminants by June 1989 and a corresponding requirement to add 25 more contaminants every three years thereafter; and (3) the designation of best available technology for each of the regulated contaminants. In response to the amendments, the EPA formulated a schedule for regulating disinfection by-products, with the by-products being considered as part of the first 25 additional contaminants to be regulated. The original EPA schedule consisted of proposed maximum contaminant levels for disinfectants and disinfection by-products by September 1990, adoption of final maximum contaminant levels by September 1991, and compliance with the adopted maximum contaminant levels by March 1993 (McGuire, 1989).

As a result of the Safe Drinking Water Act amendments, it was anticipated that: the previous maximum contaminant level for total THMs would be modified by adopting individual maximum contaminant levels for each of the four THM species based on knowledge of the differing toxicological impacts of each of the species; consideration would be given to making the new maximum contaminant levels for the THMs more stringent than the maximum contaminant level for total THMs previously adopted in 1979; maximum contaminant levels would be established for a number of the additional disinfection by-products identified in finished drinking water for which data on health effects were available; maximum contaminant levels would be established for the disinfectants themselves; and the maximum contaminant levels would apply to all water systems, not just to those serving more than 10,000 people.

The Surface Water Treatment Rule, which was first proposed in 1987 (EPA, 1987a) and finalized in 1989 (EPA, 1989a), arose from growing concerns about the presence of pathogenic microorganisms in surface water supplies and the inadequacy of existing treatment and disinfection practices to properly remove and inactivate many of these organisms (Craun, 1988). Until that time, disinfection practices were directed at the removal and inactivation of enteric (i.e., intestinal) bacteria. But growing concerns about waterborne viruses such as the hepatitis A virus, cysts such as *Giardia lamblia*, bacteria such as *Legionella*, and the number of reported illnesses associated with these organisms led to the requirements that disinfection practices be adopted to ensure at least 99.9 percent (three orders of magnitude, or 3-log) removal/inactivation of *Giardia* cysts and 99.99 percent (four orders of magnitude, or 4-log) removal/inactivation of viruses. It had become clear that existing disinfection practices might not be adequate for controlling the diverse nature of microorganisms associated with infectious diseases potentially carried by municipal drinking water.

At the heart of the Surface Water Treatment Rule is the CT concept, which specifies that a sufficient concentration (C) of disinfectant must persist in the water for a satisfactory contact time (T) in order to ensure an adequate degree of inactivation. CT values were developed from experimental results for a variety of disinfectants, specifically free chlorine, chloramines, chlorine dioxide, and ozone, for a variety of solution conditions (e.g., pH and temperature), and for various degrees of inactivation of *Giardia* and viruses. CT credit of 2- to 2.5-log removal (99 to 99.7 percent) for *Giardia* and 1- to 2-log removal (90 to 99 percent) for viruses were allocated to surface water treatment systems practicing filtration, with the remaining degrees of inactivation of the two groups of organisms to be achieved by chemical disinfection.

Another key aspect of the Surface Water Treatment Rule that affected the disinfection by-product issue was the requirement that a disinfectant residual of at least 0.2 mg/L must be maintained at all times in the water entering the distribution system, and measurable disinfectant residuals must be achieved in more than 95 percent of the distribution system samples analyzed.

The Total Coliform Rule was also initially proposed in 1987 (EPA, 1987b) and finalized in 1989 (EPA, 1989b). This rule also mandated greater attention to disinfection practices, particularly with regard to maintaining the biological quality of treated water in the distribution system.

A significant impact of the disinfection requirements of both the Surface Water Treatment Rule and the Total Coliform Rule was that many of the water utilities, such as those surveyed by McGuire and Meadow (1988), which had previously modified their disinfection practices to comply with the 1979 THM regulation might not be in compliance with the CT or residual disinfectant requirements of the Surface Water Treatment Rule and the various stipulations of the Total Coliform Rule. This was especially true for those utilities that adopted chloramination for primary disinfection. Accordingly, the population served by these utilities was potentially being exposed to a greater microbial risk as a result of modified disinfection practices. Furthermore, even without the Surface Water Treatment Rule and Total Coliform Rule, most of the utilities surveyed by McGuire and Meadow indicated they would have difficulty complying with a maximum contaminant level for total THMs significantly below 50 μg/L. Clearly, given the stringent provisions of the two new rules, reduction of the maximum contaminant level to less than 50 μg/L would be expected to have an even greater economic impact than that projected by the McGuire and Meadow survey.

Hence, in developing national regulations for the control of disinfectants and disinfection by-products, the EPA must ensure that the maximum contaminant levels established for the disinfectants and disinfection by-products are consistent with the requirements of the Surface Water Treatment Rule, the impending Groundwater Disinfection Rule (EPA, 1992a), and the Total Coliform Rule in that they do not cause changes in water treatment practices that result in significant increases in risk from waterborne pathogens or from other disinfection by-products that do not get regulated at that time. Furthermore, as stated above, the new regulations would have to apply to all public water systems using disinfection, not just to those serving more than 10,000 persons.

In developing the maximum contaminant levels for disinfection by-products, EPA's approach has been to consider various THM concentrations as maximum contaminant levels, evaluate different treatment options that could meet these maximum contaminant levels, determine the costs associated with these options, and establish an appropriate maximum contaminant level for the THMs with corresponding treatment technologies to achieve compliance. The focus has continued to be directed at THMs because of the greater availability of data on occurrence, toxicological effects, and treatment control methodologies, although increasing amounts of data are being generated for the haloacetic acids in response to their perceived significance from an occurrence and health effects point of view.

Because chloroform, bromodichloromethane, bromoform, and dichloroacetic acid are tentatively classified as probable human carcinogens based on the latest toxicological information (Regli et al., 1992), the Safe Drinking Water Act essentially requires the EPA to set the maximum contaminant level goals for these compounds at zero. The Safe Drinking Water Act also requires the EPA to set the actual maximum contaminant levels as close to the maximum contaminant level goal as is technologically and economically feasible to achieve. For carcinogenic compounds that have a maximum contaminant level goal of zero, the EPA attempts to establish maximum contaminant levels at concentrations which ensure that the average individual lifetime risk of acquiring cancer through exposure to these compounds in drinking water is no more than one in ten thousand to one in a million (equivalent to a 10^{-4} to 10^{-6} risk).

For those chlorinated disinfection by-products tentatively classified as probable human carcinogens, Table 4 lists the drinking water concentrations corresponding to 10^{-4}, 10^{-5}, and 10^{-6} increased lifetime cancer risks (Regli et al., 1992), along with the reported range of occurrence. Given the measured concentrations of these disinfection by-products in drinking water, it is clear that utilities would have great difficulty complying with maximum contaminant levels for the THMs at anything more stringent than a 10^{-5} risk level, and a maximum contaminant level for dichloroacetic acid (DCAA) at anything more stringent than a 10^{-4} risk level. While it might be possible to achieve maximum contaminant levels corresponding to these risk levels by reducing the dose of chlorine for drinking water treatment, the Surface

TABLE 4 Drinking Water Concentrations Associated with Various Levels of Increased Lifetime Cancer Risk

Disinfection By-Product	Concentration (μg/L)			Range of Occurrence (μg/L)
	10^{-4}	10^{-5}	10^{-6}	
Bromodichloromethane	100	10	1	0–100
Bromoform	400	40	4	0–50
Chloroform	600	60	6	0–340
Dichloroacetic acid	10	1	0.1	0–80

SOURCE: Regli et al. (1992).

Water Treatment Rule prevents this from happening by requiring strict adherence to CT values and disinfectant residuals that would ensure compliance with the specified levels of pathogen inactivation and the corresponding reduction in microbial risk.

Alternatively, other disinfectants could be used in place of free chlorine to meet the requirements of the Surface Water Treatment Rule, but these alternatives are not without their own health risks. For example, the use of chlorine dioxide results in the presence of chlorite (ClO_2^-) and chlorate (ClO_3^-) in the treated water (Rav-Acha et al., 1984; Werdehoff and Singer, 1987). Chlorite is formed as a by-product from the reaction of chlorine dioxide, and both chlorite and chlorate are frequently found as contaminants in chlorine dioxide feed streams (Griese et al., 1991). Chlorite has been found to be toxic in animals (Couri et al., 1982), and a maximum contaminant level goal of 0.3 mg/L has been suggested (Regli et al., 1992).

In the case of ozone, the principal by-products appear to be aldehydes such as formaldehyde, acetaldehyde, glyoxal, and methyl glyoxal (Glaze et al., 1991). There is not enough information at this time to ascertain the health risks of exposure to these chemicals in drinking water. However, in water containing bromide, ozonation can lead to the formation of undesirable levels of bromate (BrO_3^-) and brominated organic by-products such as bromoform, dibromoacetic acid, etc. (Amy and Siddiqui, 1991; Krasner et al., 1993). Bromate is also tentatively classified as a probable human carcinogen, with a 5 μg/L concentration in drinking water corresponding to a 10^{-4} individual lifetime cancer risk, and 0.5 μg/L (below most analytical detection limits) for a 10^{-5} risk (Regli et al., 1992). Therefore, while ozone is a more potent disinfectant than free chlorine, its widespread applicability may be limited by the bromide content of the source water and the corresponding formation of bromate. In addition, because ozone is relatively unstable in water, a secondary disinfectant is required in order to maintain a disinfectant residual in the distribution system.

The use of chloramines in place of free chlorine to minimize disinfection by-product formation is limited because of their poor virucidal and cysticidal properties, that is, the high CT requirements make chloramines impractical for primary disinfection in most cases. Chloramines still remain an attractive option for secondary disinfection, that is, for maintaining a persistent disinfectant residual in the distribution system.

Also, the application of alternative primary disinfectants may enhance the formation of some by-products during secondary disinfection. For example, ozonation leads to high concentrations of chloral hydrate when the secondary disinfectant is free chlorine (Logsdon et al., 1992; McKnight and Reckhow, 1992).

Some additional confounding factors involve the emergence of *cryptosporidiosis* as a major waterborne disease; several outbreaks have occurred within the past five years (Rose, 1988). Although they were not considered in the development of the Surface Water Treatment Rule, *Cryptosporidium* cysts have been found to be appreciably more resistant to conventional disinfectants than *Giardia*, and higher CT values may be required to inactivate this organism. Such CT criteria are still in the developmental stage. In addition, LeChevallier et al. (1991) have recently reported that concentrations of *Giardia* and *Cryptosporidium* cysts in raw drinking water may be orders of magnitude higher than originally expected. Although the infectivity of the cysts was not verified, these findings suggest that the specifications for 3-log and 4-log inactivation or removal of *Giardia* and viruses, respectively, in the Surface Water Treatment Rule may not be restrictive enough to reduce the risk of exposure to these waterborne diseases to levels that EPA deems acceptable. Accordingly, the EPA is considering adopting an "enhanced" Surface Water Treatment Rule. If this were done, increased contact times with chlorine or higher doses of chlorine would be required to provide enhanced disinfection. The result would almost certainly be an increase in the formation of halogenated by-products.

These considerations draw attention to the conundrum facing water utilities and regulators, that is, while a great deal of attention has been directed at formation of disinfection by-products and reducing the chemical risks associated with these by-products, insufficient attention has been focused on the microbial risks associated with modified disinfection practices and the presence of disease-causing and disinfection-resistant organisms in raw water supplies. From the standpoint of reducing microbial risk from these infectious agents, greater reliance on chlorine and other chemical disinfectants might be desirable despite the increased formation of disinfection by-products and their attendant chemical risks.

All of the above confounding factors are considered, to some degree, in EPA's Disinfection By-Product Risk Assessment Model (Regli et al., 1992). The EPA first proposed this model in November 1990 as an innovative approach to balance the risk between exposure to microbial hazards and exposure to disinfectants and disinfection by-products. The approach involves (1) identification of candidate best available technologies for controlling both disinfection by-products and pathogens; (2) determination of the degree to which these technologies are consistent with the criteria of the Surface Water Treatment Rule and the Total Coliform Rule, and with the potential criteria of an enhanced Surface Water Treatment Rule; (3) evaluation of the predicted performance of these candidate best available technologies in a variety of source water qualities (e.g., total organic carbon, bromide concentrations); (4) consideration of candidate maximum contaminant levels for THMs and haloacetic acids based upon predicted levels that can be achieved by the candidate best available technologies for a reasonable number of water systems; (5) consideration of potential net changes in risk from exposure to both disinfection by-products and pathogens; and (6) the costs to implement such changes.

EPA is still developing their risk assessment model for disinfection by-products. Initial predictions indicated that there would be a dramatic increase in the incidence of waterborne disease if more stringent maximum contaminant levels for THMs and other disinfection by-products were established, but there is a vast amount of uncertainty associated with the model, deriving primarily from inadequate scientific (primarily toxicological and microbiological), technological, and cost information. While the risk assessment modeling approach is a laudable one, it represents a rather ambitious undertaking, given the lack of supporting documentation that is needed for incorporation into the model.

In fact, it is the lack of such information in the face of the confounding factors enumerated above that, in large part, has led the EPA to recommend that the disinfection by-product regulations be developed through the process of negotiated rulemaking (RegNeg). This process is an alternative to traditional rulemaking, in which representatives of all interested parties, including the EPA, meet and collectively develop the proposed rule by consensus. The process is managed by a third party that convenes the meetings and oversees the deliberations. On September 15, 1992, the EPA issued a notice of intent to proceed with a negotiated rulemaking (EPA, 1992b). The RegNeg process for disinfection by-products is under way and is scheduled to be completed by mid-1993.

DISCUSSION AND CONCLUSIONS

There can be no questioning the fact that the regulation of disinfection by-products is a difficult issue. In some ways, the regulatory process for THMs worked effectively, at least in the early stages. Scientific evidence supported the need to control THMs. THMs were found in chlorinated water and were produced as a result of the chlorination process. They induced cancers in laboratory animals. Epidemiological studies suggested that people drinking chlorinated water appeared to be subjected to a somewhat increased incidence of cancers of the urinary and digestive tracts. Accordingly, THMs were regulated. The result of the regulation was a lowering of THM levels in drinking water at a relatively modest cost. Scientific findings led to a regulatory decision that produced a favorable outcome at little cost to society, although there is some question whether the effectiveness of disinfection was adversely affected. All of the above was accomplished over a period of about 10 years.

It has been the attempt to fine-tune and expand the original objective, and subsequent recognition that control of the chemical risks associated with disinfection by-products may increase risks associated with disease-causing microorganisms, that has led to the difficulties described in this paper.

New analytical techniques have allowed many more contaminants in drinking water to be identified. Some of these contaminants are present in the raw water, and a number are produced by chemicals added to purify the water. Some, like the THMs and haloacetic acids, are present in concentrations as high as 50–100 µg/L. Most are present at what would be called trace concentrations, that is, less than 10 µg/L. Despite the survey by McGuire et al. (1989), there is no comprehensive data base profiling the nationwide occurrence of disinfection by-products in chlorinated drinking water.

Among the disinfection by-products, halogenated by-products have been the easiest to identify and quantify analytically, although only about 50 percent of the mass of these halogenated by-products (on a halogen-equivalent basis) has been accounted for. Little information is available on the nature of the remaining 50 percent, or the health effects associated with this material. Identification of by-products resulting from the chlorination of water continues to be an active area of analytical research.

In addition, the focus of analytical activity on by-products of chlorination does not mean that alternative oxidants and disinfectants do not generate by-products too. Analytical techniques have not advanced to the point where the polar compounds that are invariably produced by these alternative chemical additives such as ozone and chlorine dioxide can be reliably measured. Identification of some of these other by-products is just beginning to occur, but it is likely to be a long time before a mass balance on the organic carbon content of finished drinking water can be attempted.

Assessment of the health effects of these trace contaminants is probably the major source of scientific uncertainty. Epidemiologically, it cannot yet be concluded that there is a significant cancer risk in drinking chlorinated water. Confident quantification of this risk requires additional study. Toxicologically, it can be concluded that some of these disinfection by-products cause cancer in laboratory animals when administered in large doses, but questions arise when attempts are made to extrapolate these results to the low exposures associated with drinking water. The scientific basis for making such low dose extrapolations is still debatable, yet the need is great if reliable risk estimates are to be made.

Technologically, it is widely accepted that the formation of THMs and other chlorinated by-products can be lowered by decreasing the use of free chlorine. However, requirements for reliable disinfection of drinking water to control waterborne pathogens are deservedly more stringent than ever before, and likely to become even more stringent. There is a fear among many that lower maximum contaminant levels for THMs and other disinfection by-products may compromise one of the fundamental objectives of water treatment, that is, disinfection. If disinfectants other than free chlorine were to be used, there are uncertainties with regard to the public health impact of some of these alternatives, particularly in bromide-containing waters. From a regulatory standpoint, it would not be desirable to establish regulations that would essentially limit the use of free chlorine and force utilities to use alternative disinfectants whose public health risk is more uncertain at this time because of analytical deficiencies.

Alternative control options involve the use of technologies that would remove disinfection by-product precursors (natural organic matter) from the water before a disinfectant is added. Technologies for achieving this objective consist of enhanced coagulation, granular activated carbon adsorption, and membrane filtration. The first of these is not effective in all waters, for reasons that are not yet known. Research on this subject is continuing. Using granular activated carbon adsorption to control disinfection by-products has been demonstrated to be a relatively expensive process for most waters, and adoption of maximum contaminant levels for disinfection by-products that are based on the use of granular activated carbon as the best available technology is likely to have a major economic impact on society. Opponents of such a requirement cite the high cost of this technology for a questionable societal benefit. Membrane filtration is a relatively new development in water treatment technology. A number of technical issues must be addressed before this technology can be implemented on a large scale. These include control of membrane fouling, ultimate disposal of the concentrated waste from the process, and efficiency of product recovery. In addition, like granular activated carbon, membrane filtration has a relatively high price tag.

EPA's proposal to use a Disinfection By-Product Risk Assessment Model to aid them in the rulemaking process is a logical approach in view of the complexities involved in the disinfection/disinfection by-products issue. Unfortunately, the scientific data base needed to support such an approach at this time is weak. Either more research is needed to provide the missing pieces, and there are a significant number of missing pieces as discussed in this paper, or a compromise position will have to be adopted, as was the case in 1979. The negotiated rulemaking process that is under way is likely to lead to such a compromise position for the short term. The long-term regulation of disinfection by-products must await additional scientific findings.

REFERENCES

Amy, G.L., and M. S. Siddiqui. 1991. Ozone-bromide interactions in water treatment. Pp. 807–824 in Proceedings of the American Water Works Association Annual Conference. Denver, Colo.: American Water Works Association.

Babcock, D. B., and P. C. Singer. 1979. Chlorination and coagulation of humic and fulvic acids. Journal of the American Water Works Association 71(3):149.

Bellar, T. A., J. J. Lichtenberg, and R. C. Kroner. 1974. The occurrence of organohalides in chlorinated drinking water. Journal of the American Water Works Association 66:703.

Bove, F. J., M. C. Fulcomer, J. B. Klotz, and others. 1992a. Report on Phase IV-A: public drinking water contamination and birthweight, fetal deaths, and birth defects; a cross-sectional study. Trenton, N.J.: New Jersey Department of Health.

Bove, F. J., M. C. Fulcomer, J. B. Klotz, and others. 1992b. Report on Phase IVB: Public drinking water contamination and birthweight, and selected birth defects; a case-control study. Trenton, N.J.: New Jersey Department of Health.

Brass, H. J., M. A. Feige, T. Halloran, J. W. Mello, D. Munch, and R. F. Thomas. 1977. The national organic monitoring survey: Samplings and analyses for purgeable organic compounds. Pp. 393–416 in Drinking Water Quality Enhancement Through Source Protection, R. B. Pojasek, ed. Ann Arbor, Mich.: Ann Arbor Science Publishers.

Brodtmann, N. V., W. E. Koffskey, and J. DeMarco. 1980. Studies on the use of combined chlorine (monochloramine) as a primary disinfectant of drinking water. Pp. 777–788 in Water Chlorination: Environmental Impact and Health Effects, Vol III, R. L. Jolley et al., eds. Ann Arbor, Mich.: Ann Arbor Science Publishers.

Bull, R. J., and F. C. Kopfler. 1991. Health effects of disinfectants and disinfection by-products. Denver, Colo.: American Water Works Association Research Foundation.

Bull, R. J., I. M. Sanchez, M. A. Nelson, J. L. Larson, and A. D. Lansing. 1990. Liver tumor induction in B6C3F1 mice by dichloroacetate and trichloroacetate. Toxicology 63:341–359.

Christman, R. F., D. L. Norwood, D. S. Millington, J. Donald Johnson, and A. A. Stevens. 1983. Identity and yields of major halogenated products of aquatic fulvic acid chlorination. Environmental Science and Technology 17(10):625–628.

Coleman, W. E., J. W. Munch, W. H. Kaylor, R. P. Streicher, H. P. Ringhand, and J. R. Meier. 1984. Gas chromatography/mass spectroscopy analysis of mutagenic extracts of aqueous chlorinated humic acid: A comparison of the by-products of drinking water contaminants. Environmental Science and Technology 18(9):674.

Couri, D., M. S. Abdel-Rahman, and R. J. Bull. 1982. Toxicological effects of chlorine dioxide, chlorite, and chlorate. Environmental Health Perspectives 46:13.

Craun, G. F. 1988. Surface water supplies and health. Journal of the American Water Works Association 80(2):40.

Craun, G. F. 1991. Epidemiological studies of organic micropollutants in drinking water. In The Handbook of Environmental Chemistry, O. Hutzinger, ed. Berlin: Springer Verlag.

Duke, D. T., J. W. Siria, B. D. Burton, and D. W. Amundsen. 1980. Control of trihalomethanes in drinking water. Journal of the American Water Works Association 72(8):470.

162 PHILIP C. SINGER

Glaze, W. H., H. S. Weinberg, S. W. Krasner, and M. J. Sclimenti. 1991. Trends in aldehyde formation and removal through plants using ozonation and biological active filters. Pp. 913–943 in Proceedings of the American Water Works Association Annual Conference. Denver, Colo.: American Water Works Association.

Griese, M. H., K. Hauser, M. Berkemeier, and G. Gordon. 1991. Using reducing agents to eliminate chlorine dioxide and chlorite ion residuals in drinking water. Journal of the American Water Works Association 83(5):56.

Herren-Freund, S. L., M. A. Pereira, M. D. Khoury, and G. Olson. 1987. The carcinogenicity of trichloroethylene and its metabolites, trichloroacetic acid and dichloroacetic acid, in mouse liver. Toxicology and Applied Pharmacology 90:183.

IJsselmuiden, C. B., C. Gaydos, B. Feighner, W. L. Novakoski, D. Serwadda, L. H. Caris, D. Viahov, and F. W. Comstock. 1992. Cancer of the pancreas and drinking water: A population-based case-control study in Washington County, Maryland. American Journal of Epidemiology 136(7):836.

Johnson, D. E., and S. J. Randtke. 1983. Removing non-volatile organic chlorine and its precursors by coagulation and softening. Journal of the American Water Works Association 75(5):249.

Jorgenson, T. A., E. F. Meierhenry, C. J. Rushbrook, R. D. Bull, and M. Robinson. 1985. Carcinogenicity of chloroform in drinking water to male Osborne-Mendel rats and female B6C3F1 mice. Fundamentals of Applied Toxicology 5:760–769.

Kavanaugh, M. C. 1978. Modified coagulation for improved removal of trihalomethane precursors. Journal of the American Water Works Association 70(11):613.

Krasner, S. W., M. J. McGuire, J. J. Jacangelo, N. L. Patania, K. M. Reagan, and E. M. Aieta. 1989. The occurrence of disinfection by-products in U.S. drinking water. Journal of the American Water Works Association 81(8):41.

Krasner, S. W., W. H. Glaze, H. S. Weinberg, P. A. Daniel, and I. N. Najm. 1993. Formation and control of bromate during ozonation of waters containing bromide. Journal of the American Water Works Association 85(1):73.

Kronberg, L., B. Holmbom, M. Reunanen, and L. Tikkanen. 1988. Identification and quantification of the Ames mutagenic compound 3-chloro-4-(dichloromethyl)-5-hydroxy-2(5H)-furanone and of its geometric isomer (E)-2-chloro-3-(dichloromethyl)-4-oxybutenoic acid in chlorine-treated humic water and drinking water extracts. Environmental Science and Technology 22(9):1097.

Lange, A. L., and E. Kawczynski. 1978. Controlling organics: The Contra Costa County Water District experience. Journal of the American Water Works Association 70(11):63.

LeChevallier, M. W., D. N. Norton, and R. G. Lee. 1991. Occurrence of Giardia and Cryptosporidium in surface water supplies. Applied Environmental Microbiology 57:2610.

Logsdon, G. S., S. Foellmi, B. Long, R. Dawson, M. Ferguson, and D. Neden. 1992. Filtration pilot plant studies for Greater Vancouver's water supply. In Proceedings of the American Water Works Association Annual Conference. Denver, Colo.: American Water Works Association.

McGuire, M. J. 1989. Preparing for the disinfection by-products rule: A water industry status report. Journal of the American Water Works Association 81(8):35.

McGuire, M. J., and R. G. Meadow. 1988. AWWARF trihalomethane survey. Journal of the American Water Works Association 80(1):61.

McGuire, M. J., S. W. Krasner, K. M. Reagan, et al. 1989. Disinfection by-products in United States drinking waters. Washington, D.C.: U.S. Environmental Protection Agency.

McKnight, A., and D. A. Reckhow. 1992. Reactions of ozonation by-products with chlorine and chloramines. In Proceedings of the American Water Works Association Annual Conference. Denver, Colo.: American Water Works Association.

Miller, J. W., and P. C. Uden. 1983. Characterization of non-volatile aqueous chlorination products of humic substances. Environmental Science and Technology 17(3):150.

Montgomery, J. M., Consulting Engineers Inc. 1992. Effect of coagulation and ozonation on the formation of disinfection by-products. Denver, Colo.: American Water Works Association.

Morris, R. D., A. M. Audet, I. F. Angelillo, T. C. Chalmers, and F. Mosteller. 1992. Chlorination, chlorination by-products, and cancer: A meta-analysis. American Journal of Public Health 82(7):955.

National Cancer Institute. 1976. Report on the carcinogenesis bioassay of chloroform. NTIS PB-264-018. Bethesda, Md.: National Cancer Institute.

National Research Council. 1980. Drinking Water and Health, Vol. 3. Report of the Safe Drinking Water Committee. Washington, D.C.: National Academy Press.

National Research Council. 1987. Drinking Water and Health: Disinfectants and Disinfectant By-Products, Vol. 7. Washington, D.C.: National Academy Press.

National Toxicology Program. 1986. Toxicology and carcinogenesis studies of bromodichloromethane in F344/N rats and B6C3F1 mice. Technical Report Series No. 321, NIH Publication No. 88-2537, CAS No. 75-27-4. Washington, D.C.: U.S. Department of Health and Human Services.

National Toxicology Program. 1989. Toxicology and carcinogenesis studies of tribromomethane (bromoform) in F344/N rats and B6C3F1 mice. Technical Report Series No. 350, NIH Publication No. 89-2805, CAS No. 75-25-2. Washington, D.C.: U.S. Department of Health and Human Services.

Norman, T. S., L. L. Harms, and R. W. Looyenga. 1980. The use of chloramines to prevent trihalomethane formation. Journal of the American Water Works Association 72(3):176.

Oliver, B. G. 1983. Dihaloacetonitriles in drinking water: Algae and fulvic acid as precursors. Environmental Science and Technology 17(2):80.

Quimby, B. D., M. F. Delaney, P. C. Uden, and R. M. Barnes. 1980. Determination of the aqueous chlorination products of humic substances by gas chromatography with microwave emission detection. Analytical Chemistry 52:259.

Rav-Acha, C., A. Serri, E. Choshen, and B. Limoni. 1984. Disinfection of drinking water rich in bromide with chlorine and chlorine dioxide, while minimizing the formation of undesirable by-products. Water Science and Technology 17:611.

Reckhow, D. A., and P. C. Singer. 1984. The removal of organic halide precursors by preozonation and alum coagulation. Journal of the American Water Works Association 76(4):151.

Regli, S., J. E. Cromwell, X. Zhang, A. B. Gelderloos, W. D. Grubbs, F. Letkiewicz, and B. A. Macler. 1992. Framework for decision making: An EPA perspective. EPA 811-R-92-005, August. Washington, D.C.: U.S. Environmental Protection Agency.

Rook, J. J. 1974. Formation of haloforms during chlorination of natural waters. Water Treatment and Examination 23:234.

Rook, J. J. 1977. Chlorination reactions of fulvic acids in natural waters. Environmental Science and Technology 11(5):478.

Rose, J. B. 1988. Occurrence and significance of Cryptosporidium in water. Journal of the American Water Works Association 80(2):53.

Semmens, M., and T. Field. 1980. Coagulation: Experiences in organics removal. Journal of the American Water Works Association 72(8):476.

Singer, P. C., and S. D. Chang. 1989. Correlations between trihalomethanes and total organic halides formed during drinking water treatment. Journal of the American Water Works Association 81(8):61.

Singer, P. C., A. Obolensky, and A. D. Greiner. 1992. Relationships among disinfection by-products in chlorinated drinking waters. In Proceedings of the Water Quality Technology Conference. Denver, Colo.: American Water Works Association.

Standard Methods for the Examination of Water and Wastewater. 1985. Sixteenth edition. American Public Health Association, Washington, D.C.

Stevens, A. A., L. A. Moore, C. J. Slocum, B. L. Smith, D. R. Seeger, and J. C. Ireland. 1989. Chlorinated humic acid mixtures: Criteria for detection of disinfection by-products in drinking water. Chapter 39 in Aquatic Humic Substances: Influence on Fate and Transport of Pollutants, I. H. Suffett and P. MacCarthy, eds. Advances in Chemistry Series 219. Washington, D.C.: American Chemical Society.

Symons, J. M., T. A. Bellar, J. K. Carswell, J. DeMarco, K. L. Kropp, G. G. Robeck, D. R. Seeger, C. J. Slocum, B. L. Smith, and A. A. Stevens. 1975. National organics reconnaissance survey for halogenated organics in drinking water. Journal of the American Water Works Association 67:634.

Symons, J. M., A. A. Stevens, R. M. Clark, E. E. Geldreich, O. T. Love, Jr., and J. DeMarco. 1981. Treatment Techniques for Controlling Trihalomethanes in Drinking Water. EPA-600/2-81-156, September. Cincinnati, Ohio: U.S. Enviromental Protection Agency.

Trehy, M. L., and T. I. Bieber. 1981. Detection, identification and quantitative analysis of dihaloacetonitriles in chlorinated drinking water. Pp. 941–975 in Advances in the Identification and Analysis of Organic Pollutants in Water, Vol. 2., L. H. Keith, ed. Ann Arbor, Mich.: Ann Arbor Science Publishers.

U.S. Environmental Protection Agency. 1976. Organic chemical contaminants: control options in drinking water. Federal Register 41(136):28991, July 14, 1976.

U.S. Environmental Protection Agency. 1978. Control of organic contaminants in drinking water. Federal Register 43(28):5756, February 9, 1978.

U.S. Environmental Protection Agency. 1979. National interim primary drinking water regulations; control of trihalomethanes in drinking water. Federal Register 44(231):68624, November 29, 1979.

U.S. Environmental Protection Agency. 1983. National interim primary drinking water regulations; trihalomethanes. Federal Register 48(40):8406, February 28, 1983.

U.S. Environmental Protection Agency. 1987a. National primary drinking water regulations; filtration and disinfection; turbidity, Giardia lamblia, viruses, Legionella, and heterotrophic bacteria. Federal Register 52(212):42173, November 3, 1987.

U.S. Environmental Protection Agency. 1987b. Drinking water; national primary drinking water regulations; total coliforms. Federal Register 52(212):42224, November 3, 1987.

U.S. Environmental Protection Agency. 1989a. National primary drinking water regulations; filtration and disinfection; turbidity, Giardia lamblia, viruses, Legionella, and heterotrophic bacteria. Federal Register 54(124):27486, June 29, 1989.

U.S. Environmental Protection Agency. 1989b. Drinking water; national primary drinking water regulations; total coliforms. Federal Register 54(212):27544, June 29, 1989.

U.S. Environmental Protection Agency. 1992a. Draft groundwater disinfection rule. Federal Register 57(148):33960, July 31, 1992.

U.S. Environmental Protection Agency. 1992b. Intent to form an advisory committee to negotiate the disinfection byproducts rule and announcement of public meeting. Federal Register 67(179):42533, September 15, 1992.

Werdehoff, K. S., and P. C. Singer. 1987. Chlorine dioxide effects on THMFP, TOXFP, and the formation of inorganic by-products. Journal of the American Water Works Association 79(9):107.

World Health Organization. 1991. Chlorinated drinking-water; chlorination by-products; some other halogenated compounds; cobalt and cobalt compounds. In IARC Monographs on the Evaluation of Carcinogenic Risks to Humans, Vol. 52. International Agency for Research in Cancer, Lyon.

Young, J. S., and P. C. Singer. 1979. Chloroform in public water supplies: a case study. Journal of the American Water Works Association 71(2):87.

Keeping Pace with Science and Engineering. 1993.
Pp. 165–188. Washington, DC: National Academy Press.

Acid Deposition

James L. Regens

Attempts to link scientific, technical, and economic information to decisions affecting the public sector have become a significant as well as
contentious component of the policymaking process in the United States in
recent decades. Even when high-quality information is available, barriers
may impede its timely communication and use. Almost inevitably, when
public policy choices are grounded heavily in scientific, technical, or economic data, much of the debate among contending sides involves conflict
over whose information will become the more credible, and persuasive, in
the political arena. The contention involved in accommodating scientific,
technical, and economic information in the decision-making process also
reflects the fact that our understanding of risks, benefits, and costs is not
static but continuously evolves. Thus, it is always possible for new information to emerge that calls into question the previously accepted scientific,
technical, or economic bases of regulatory choices.

In a number of instances, environmental statutes incorporate explicit
procedures for revising regulations to accommodate new data. The Clean
Air Act Amendments of 1990 (P.L. 101-549), for example, provide a process for periodically revising the National Ambient Air Quality Standards;
there are comparable mechanisms within the regulatory system that make it
possible for administrative agencies to accommodate changes in existing
knowledge after regulations are promulgated. Given concerns about the
role that scientific, technical, or economic understanding plays in the rulemaking process, it is useful to delineate the ways in which changing information similarly affects congressional deliberations about new legislation.

The acid rain controversy provides an excellent case study of the potential for changing information to guide policy choices during congressional debates as well as the limitations of accommodation.[1]

The acid deposition control program authorized by Title IV of the Clean Air Act Amendments of 1990 signaled the end to more than a decade of acrimonious debate. However, before these regulations were passed acid rain was one of the most prominent, complex, and divisive environmental research and policy issues of the 1980s (see Regens and Rycroft, 1988). Both scientific knowledge and governmental policy were controversial. The controversy over science centered on how much information was needed to determine if acid rain was a threat and whether it could be prevented or mitigated. The policy controversy involved the appropriateness of alternative responses to such a threat. A tremendous amount of research was conducted and the disagreement among interested parties was intense— especially the policy implications of research findings.

What lessons can we learn from that experience that will improve the use and effectiveness of scientific and economic information in congressional policy debates? Can the lessons improve the effectiveness of large-scale interagency research programs as a mechanism for generating such information?[2] The answers may be valuable in shaping timely and prudent responses to other major atmospheric pollution issues, such as global climate change or stratospheric ozone depletion.

OVERVIEW OF EXISTING INFORMATION

Definition and Origin

As a working definition, acid deposition, or acid rain as it is more commonly called, refers to the processes by which acidic substances, which are largely of man-made origin, are deposited from the atmosphere into ecosystems in precipitation or as fine dry particles. Acidity is measured on a logarithmic scale (pH) of 1 to 7, with 7 being neutral and acidity increasing as the numbers decrease toward 1.[3] As shown in Figure 1, all forms of precipitation—rain, snow, sleet, hail, fog, or mist—that have a pH value equal to or less than 5.6 typically are classified as acid rain.[4] In fact, however, a review of deposition data in the existing literature suggests that the global average pH of precipitation in remote regions of the world is closer to 5.0, which appears to be a more appropriate cutoff point for "clean" rain. The "natural" value of precipitation pH probably varies from region to region depending on the climatology, local ecosystems, and other factors.

Although the simplicity of the term *acid rain* conveys the image of an easily understood phenomenon, Figure 2 shows how the problem involves complex and varied chemical, meteorological, and physical interactions (see

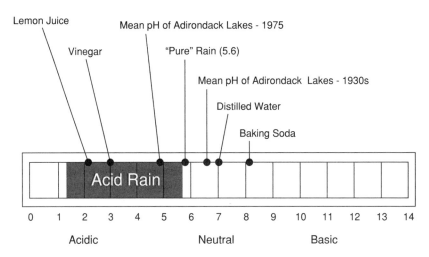

FIGURE 1   The pH scale.

National Acid Precipitation Assessment Program [NAPAP], 1991a; National Research Council [NRC], 1983, 1986). Interestingly enough, the first scientific studies attempting to delineate the processes producing acid rain date back to the late 1800s (see Figure 3). Robert Angus Smith's pioneering studies of precipitation chemistry and its effects introduced the world to the term acid rain. Drawing on data measuring the chemistry of rain in England, Scotland, and Germany, Smith (1872) demonstrated that variation in regional factors such as wind trajectories, the amount and frequency of precipitation, decomposition of organic matter, proximity to seacoasts, and coal use influenced sulfate concentrations in rain. However, Smith's research was ignored for almost a century by both the scientific and policy communities.

Research Efforts

Serious interest in acid rain as a topic for scientific inquiry did not emerge until the early 1970s, and research throughout that decade emphasized the contribution of sulfur compounds to acidification (see Cowling, 1982). Research that linked air mass trajectories to changes in precipitation chemistry (see Oden, 1968) provided the initial basis for concluding that acid deposition is caused by human activities and that it is a regional-scale environmental problem with long-term adverse consequences. The initial series of follow-on studies, primarily conducted by researchers in western Europe, concentrated on (1) delineating the effects of acid rain on aquatic

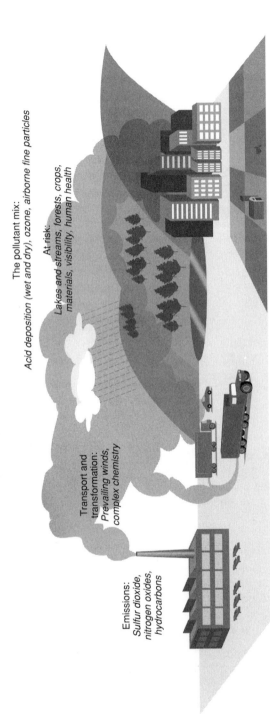

The pollutant mix:
Acid deposition (wet and dry), ozone, airborne fine particles

At risk:
Lakes and streams, forests, crops, materials, visibility, human health

Transport and transformation:
Prevailing winds, complex chemistry

Emissions:
Sulfur dioxide, nitrogen oxides, hydrocarbons

FIGURE 2   Schematic view of the acid deposition problem.   SOURCE: Office of Technology Assessment (1984).

Science and Engineering

Policy and Regulation

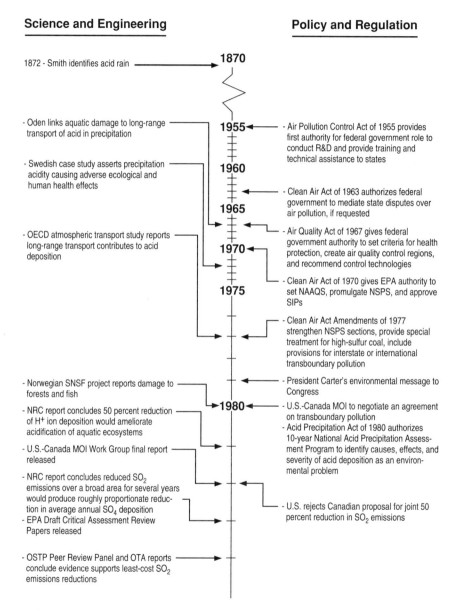

FIGURE 3 Timeline of significant scientific, technical, and regulatory developments in acid rain. (*Figure continues on next page.*)

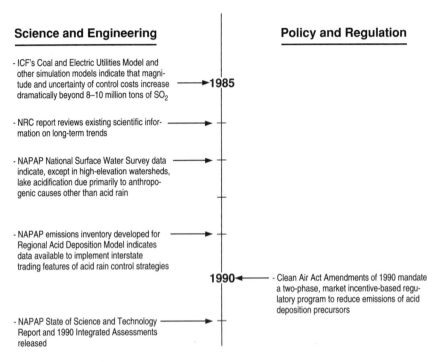

Science and Engineering

- ICF's Coal and Electric Utilities Model and other simulation models indicate that magnitude and uncertainty of control costs increase dramatically beyond 8–10 million tons of SO$_2$

- NRC report reviews existing scientific information on long-term trends

- NAPAP National Surface Water Survey data indicate, except in high-elevation watersheds, lake acidification due primarily to anthropogenic causes other than acid rain

- NAPAP emissions inventory developed for Regional Acid Deposition Model indicates data available to implement interstate trading features of acid rain control strategies

- NAPAP State of Science and Technology Report and 1990 Integrated Assessments released

1985

1990

Policy and Regulation

- Clean Air Act Amendments of 1990 mandate a two-phase, market incentive-based regulatory program to reduce emissions of acid deposition precursors

FIGURE 3 *Continued*

ecosystems and forests, and (2) using atmospheric transport models to estimate source-receptor relationships.

The Swedish case study prepared for the 1972 United Nations Conference on the Human Environment in Stockholm asserted that acid rain was due primarily to sulfur dioxide (SO$_2$) emissions from man-made sources—predominately coal-fired, steam electric power plants and industrial facilities—and that it adversely affected ecosystems and human health (Swedish Ministry of Foreign Affairs and Swedish Ministry of Agriculture, 1972). The Norwegian Interdisciplinary Research Program, commonly referred to as the SNSF project, was conducted from 1972 to 1980 and focused on establishing effects on forests and fish. Like its Swedish counterpart, the SNSF project found conclusive evidence of chemical and biological changes, including fish kills and reproductive failure, in lakes and streams that had limited capacity to neutralize acidic inputs (Overrein et al., 1980). Another major research project under the auspices of the Organization for Economic Cooperation and Development (OECD) concluded that the acid deposition occurring over almost all of northwestern Europe was due to transboundary as well as local emissions of SO$_2$. Unfortunately, because of serious problems with the reliability of the available data bases on national emissions coupled with the quality of existing atmospheric transport models, the findings of the OECD study have an extremely large range of uncertainty (±50 percent) for individual receptor sites (OECD, 1977).

By the mid-1970s, papers noting declining pH levels and speculating about the possible impact of acidification due to sulfate ($SO_4^=$) deposition on aquatic resources stimulated similar concerns about the environmental consequences of acid deposition in the United States and Canada (see Beamish and Harvey, 1972; Cogbill and Likens, 1974; National Research Council of Canada, 1981). Responding to such findings, in 1978 the United States and Canada established a Bilateral Research Consultation Group on the Long-Range Transport of Air Pollutants to coordinate the exchange of scientific information about acid rain. In 1980 the two governments took further steps to cooperate in the exchange of scientific, technical, and economic information about acid deposition when a set of three bilateral work groups composed of government experts in each of these areas was created to support negotiations under the U.S.-Canada Memorandum of Intent (MOI) concerning Transboundary Air Pollution (U.S. Department of State, 1981). The work groups were on impact assessment; atmospheric modeling; and emissions, costs, and engineering assessment, respectively.

Disagreements within the work groups over dose-response functions (i.e., the relations between the amount of a substance received and the effects it produces) for damage attributable to acid deposition as well as over reduction targets were reflected in the final summary reports of the technical work groups (see U.S.-Canada Work Groups 1, 2, and 3B, 1982). In fact, because of the policy implications of the findings presented in those documents, each country conducted its own external peer review. The U.S. review was conducted under the auspices of the White House's Office of Science and Technology Policy (OSTP), and both reviews ultimately concluded that the then available information supported selective reductions in SO_2 emissions (see Nierenberg et al., 1984). A separate evaluation of available scientific, technical, and economic information conducted by the Office of Technology Assessment (U.S. Congress, OTA, 1984) as well as an earlier report prepared by the National Research Council (NRC, 1981) reached the same conclusion. The public release of these three reports with prestigious scientific imprimaturs was a major reason the Reagan administration felt compelled to initiate limited planning for a national strategy to reduce acid deposition and shifted away, at least symbolically, from exclusive reliance on its requirement for further research (Regens and Rycroft, 1988).

Starting in the early 1980s and continuing throughout that decade, research on acid deposition expanded dramatically in scope and funding level. Unlike the pioneering studies of the 1970s, which focused almost exclusively on sulfate deposition, what legitimately can be termed "second generation" research addressed the contributions of other precursor pollutants, especially nitrogen oxides (NO_x) and volatile organic compounds (VOCs). There was also extensive research on control technologies and the effects of mitigation strategies. Four major efforts are worth noting: (1) the series of reports addressing ecological effects and atmospheric processes prepared by the National Research Council; (2) the studies on the effects of adding lime to aquatic ecosystems studies by Living Lakes, a nonprofit research group funded primarily by the electric utility industry to assess aquatic mitigation options; (3) the acid deposition research program conducted by the Electric Power Research Institute (EPRI); and (4) the National Acid Precipitation Assessment Program. Each contributed to developing and synthesizing information about the nature and extent of the adverse effects associated with acid deposition as well as the potential scientific, technical, or economic efficacy of responses to the problem.

National Acid Precipitation Assessment Program

Because of its scale (total expenditures were approximately $530 million expressed in current dollars), it is useful to describe briefly the federal government's efforts to develop and synthesize information about acid deposition under the umbrella of the National Acid Precipitation Assessment Program. The Acid Precipitation Act of 1980 (P.L. 96-294, Title VII of the Energy Security Act of 1980) authorized a 10-year research effort, commonly referred to as NAPAP, to assess possible damage to natural ecosystems, agriculture, materials, and human health. The act established an Interagency Task Force on Acid Precipitation that consisted of representatives from 12 agencies, directors of 4 national laboratories, and four presidential appointees to plan and coordinate NAPAP's implementation of a comprehensive research plan. The plan was to:

• Identify the sources of atmospheric emissions contributing to acid precipitation.
• Conduct a nationwide long-term monitoring network to detect and measure levels of acid precipitation.
• Delineate the processes by which atmospheric emissions are transformed into acid precipitation.
• Develop and apply atmospheric transport models for predicting long-range transport of substances causing acidic precipitation.
• Define geographic areas at risk by monitoring deposition to identify sensitive areas.
• Build data bases of water and soil chemistry in receptor areas.
• Develop dose-response functions for effects.
• Prepare integrated assessments of (1) the environmental impacts caused

by acidic precipitation on crops, forests, fisheries, recreational and aesthetic resources, and structures; and (2) alternative technologies to prevent or ameliorate harmful effects.

According to its original operating plan, NAPAP was to provide an initial damage assessment with preliminary estimates of acid rain impacts by 1985, focusing on the northeastern United States, and two additional integrated assessments in 1987 and 1989 to support policymaking. In addition, NAPAP's legislative mandate called for providing annual reports to the President and Congress on the status and significance of the continuing research effort as well as recommending specific policy actions to deal with acid rain. The research agenda and the schedule for the assessments and other reporting requirements outlined above were extremely ambitious and required a high level of coordination across the participating agencies.

Results

After more than two decades of focused research, it seems appropriate to ask what scientific, technical, or economic insights have been gained. First, the overall chemistry of acid-forming compounds, sources of precursor pollutants, and the importance of man-made emissions of NO_x and VOCs in addition to SO_2 as acid rain precursors are reasonably well defined. While SO_2 emissions come primarily from large point sources concentrated in a relatively small number of locales, emission of volatile organic compounds (VOC) and oxides of nitrogen (NO_x) are more evenly distributed among point and mobile sources and are more uniformly dispersed spatially among regions. Second, regional models for simulating acid deposition transport processes—especially comprehensive, process-oriented models or even relatively simple statistical models—are capable of providing consistent, fairly accurate quantitative information about source-receptor relationships when projections are averaged over yearly time periods. Third, although data are insufficient to analyze regional trends in dry deposition, time series data for wet deposition reveal that the areas of maximum deposition of acid rain in precipitation are located in the northeastern United States (see Table 1). Fourth, research on effects, including NAPAP's National Surface Water Survey, EPRI's Integrated Lake Watershed Acidification Study (ILWAS), and the Living Lakes program, demonstrated that acid deposition produces chemical and biological changes in aquatic ecosystems that have limited capacity to neutralize acids. Adding lime to surface waters or watersheds typically reverses adverse biological, chemical, and physical changes in those sensitive aquatic ecosystems. Similar evidence of direct effects on terrestrial ecosystems, materials, or human health remains inconclusive or is lacking, but there is a widespread consensus that acid deposi-

TABLE 1   Trends in Wet Deposition of Cations and Anions in the
Eastern and Western United States, 1985–1990 (in kilograms per hectare)

Year	Cations		Anions			Precipitation (cm)
	H^+	NH_{4+}	Ca^{+2}	SO_{4-}	NO_{3-}	
Eastern U.S. (79 sites)						
1985	0.40	2.36	1.55	21.33	13.08	107.2
1986	0.39	2.34	1.29	21.57	13.11	102.7
1987	0.38	2.44	1.29	20.34	12.79	100.9
1988	0.34	1.81	1.50	19.64	12.05	95.9
1989	0.38	3.09	1.44	21.45	14.07	110.0
1990	0.41	3.17	1.43	22.06	14.18	122.7
Western U.S. (44 sites)						
1985	0.06	0.96	1.23	5.25	4.09	61.5
1986	0.05	1.07	1.18	5.51	4.47	72.0
1987	0.06	1.28	1.06	5.26	4.68	62.1
1988	0.05	0.70	1.16	4.72	3.83	56.1
1989	0.04	1.35	1.18	4.73	4.44	56.0
1990	0.05	1.57	1.18	5.16	4.90	67.0

SOURCE: National Acid Deposition Program/National Trends Network data.

tion contributes to indirect effects on those resources as well as to reduced
visibility.   Fifth, economic assessment studies have evaluated options for
achieving emissions reductions under various regulatory scenarios.

OVERVIEW OF THE REGULATORY STRATEGY

Figure 3 identifies the key events in the development of the regulatory
strategy for managing the environmental consequences of acid deposition.
The first air pollution control legislation adopted at the national level in the
United States was passed in 1955 (Air Pollution Control Act of 1955, P.L.
84-159).   The federal government's role was limited to research, training,
and technical aid to state governments.   Eight years later, the Clean Air Act
of 1963 (CAA, P.L. 88-206) was passed.   While continuing the previous
emphasis on federal support of scientific and technical advice to the states,
the CAA also expanded the federal government's role as a facilitator of
intermunicipal and interstate air quality efforts.   Building on the 1963 CAA,
the federal government's preeminence in air pollution control was enhanced
by the Air Quality Act of 1967 (P.L. 90-148), which authorized the federal
government to establish criteria for health protection, designate air quality
control regions, and recommend specific control technologies for pollution

abatement, although the states retained standard-setting and enforcement responsibility.

The Clean Air Act Amendments of 1970 (P.L. 91-604) gave the federal government additional authority to deal with ambient air quality problems by shifting responsibility for the standard-setting process from the states to the federal government. Three key features of the 1970 amendments are relevant to the evolution of the regulatory strategy for managing acid deposition:

• The federal government was authorized to promulgate uniform national ambient air quality standards (NAAQS) for certain pollutants.
• The federal government was authorized to promulgate uniform new source performance standards (NSPS) limiting emissions from new point sources of pollution.
• The states were required to formulate state implementation plans (SIPs), which were subject to EPA review, to attain NAAQS compliance.

Those provisions potentially provided a framework for limiting emissions of acid deposition precursors from new sources under NSPS and existing sources under SIPs, but not for regulating acid deposition per se.

In 1977 the CAA was reauthorized with a new series of amendments that had implications for acid deposition control (Clean Air Amendments of 1977, P.L. 95-95). Reflecting emerging transboundary concerns, several sections were included in an attempt to address interstate and international effects using the existing SIP process. Section 126 permitted states or their subdivisions to seek relief from interstate pollution under section 110(a)(2)(E), which theoretically limited emissions from one state that caused ambient concentrations in another to exceed NAAQS. Section 115 presumably provided administrative procedures for addressing transboundary air quality concerns, including noncriteria pollutants. However, the statutory language of these sections is vague and, when considered as a set, they have proved to be an ineffective remedy. Two other provisions of the 1977 amendments also are worth noting:

• State governors were authorized to mandate the use of locally mined coals to prevent severe economic disruption or unemployment.
• The EPA was required to promulgate a revised NSPS for coal-fired power plants that specified a minimum percentage reduction in SO_2 emissions based on the use of best available control technology (BACT) for continuous emission control.

The 1977 amendments essentially required flue gas desulfurization (FGD) for all new coal-fired, steam electric power plants, regardless of the sulfur content of the coal used as fuel.

Finding an appropriate regulatory response to acid deposition within the framework of the CAA emerged as a public policy concern in the United States by the late 1970s. In 1979, then President Carter referred to acid precipitation as a global environmental problem of the greatest importance in a message to Congress asking for expanded research and development as well as possible control measures under the CAA (Carter, 1980). The U.S.-Canada memorandum of intent to negotiate an agreement to control transboundary air pollution and the Acid Precipitation Act of 1980 (P.L. 96-294) typified the Carter administration's dual approach to the acid deposition problem: research was to be coupled with regulatory action. Thus, better information was needed, but the need to develop a more extensive knowledge base with less uncertainty was not to be used as a rationale to delay the transition from research to regulatory interventions.

Although scheduled for reauthorization once again in September 1981, revision of the CAA was delayed throughout the 1980s largely due to controversies about a regulatory regime for acid deposition. Over 70 proposals dealing with acid rain were introduced in the Congress during that period, but no legislation was enacted until late 1990. The proposals typically called for reductions ranging from 8 to 10 million tons in annual SO_2 emissions to be achieved by 1993 using traditional command-and-control approaches (see Regens, 1989; Regens and Rycroft, 1986). The Reagan administration consistently opposed any new regulatory strategy for acid deposition control and argued that legislation was premature because of the scientific uncertainties associated with its effects and the potential efficacy of any given control strategy. Without administration support, it was impossible to get legislation that would mandate a regulatory regime specifically designed to reduce emissions of precursors of acid deposition (see Regens, 1989).

Because the candidates of the two major parties in 1988 both supported an acid deposition control program, that presidential election removed executive branch opposition to the idea of reducing emissions in principle, but it did not necessarily produce agreement on a specific regulatory strategy. For example, during the campaign, George Bush expressed the desire to become the "environmental President" and promised to propose action to control precursor emissions. In late 1989, the Bush administration submitted a proposal to amend the CAA that emphasized a market incentive-based approach to regulation instead of the more traditional command-and-control approach that dominated previous proposals to manage acid deposition. By early 1990, both chambers of Congress were working actively on legislation incorporating an incentive-based strategy and, on November 15, the Clean Air Act Amendments of 1990 (P.L. 101-549) were signed by President Bush (see Bohi and Burtraw, 1991).

Title IV of the 1990 CAA amendments, the major acid rain provision, focuses emissions reduction strategy on the electric utility industry and seeks to reduce annual SO_2 emissions from coal-fired electric utility plants approximately 10 million tons by the year 2000, with four-year extensions in meeting compliance deadlines available for companies that use clean coal repowering technology. Phase I will begin in 1995 and affect 110 plants located primarily in the midwestern and eastern United States. Phase II will begin in 2000 and cap utility SO_2 emissions at 8.95 million tons per year. Nitrogen oxide emissions are targeted to be reduced by 2 million tons per year by 2000. The EPA proposed Acid Rain Rules in December 1991 that cover allowance trading, excess emissions, permits, and continuous emissions monitoring. The public comment period closed in February 1992 and the final rules were released in late 1992. Under its Phase I rules, EPA distributed 5.7 million permits, each of which allows a discharge of 1 ton of SO_2/year between 1995 and 2000, to the electric utilities operating the 110 worst polluting power plants in terms of SO_2 emissions. Firms can use their allowances to comply with the 1990 CAA, or they can clean up their operations and sell their unused permits to utilities with pollution problems, environmental groups, or speculators. The EPA also set aside some allowances to be auctioned annually as a way of creating a market, with the proceeds to be distributed among those utilities.

The Chicago Board of Trade (CBOT) held its initial sealed-bid auction for allowances on March 29, 1993. Of the 150,010 permits sold in the first public auction for prices ranging between $122 to $450 a ton, 150,000 were allocated by EPA. Utilities offered an additional 125,000 allowances, but only 10 were purchased because the power companies set minimum prices that were higher than what buyers offered. The biggest single buyer was Carolina Power & Light, which spent approximately $11.5 million to buy 85,103 allowances at prices ranging from $122 to $171 per ton. Other utilities that bought permits included American Electric Power Service, Gulf Power, Illinois Power, Kentucky Power, and Mississippi Power. One environmental group, the National Healthy Air License Exchange, also purchased allowances to reduce the number available and boost prices.

Separately, the CBOT has developed plans to hold its own quarterly auctions of emissions allowances and establish a spot market and futures market for the permits. Several utilities already have made private trades. For example, Wisconsin Power & Light has sold more than 25,000 of its 1-ton allowances to other utilities in private deals at prices ranging from $250 to $400 a ton. Over time, the market price is likely to rise as compliance deadlines approach and emission limits tighten.

LINKS BETWEEN INFORMATION
AND POLICY CHOICE

Because public policies represent responses to perceived problems, it is worth considering whether scientific, technical, or economic information developed through NAPAP or elsewhere helped shape the regulatory strategy that eventually emerged, as outlined above. Early scientific studies here and abroad, of course, brought concerns about the causes and effects of acid deposition to the forefront as an environmental policy issue. For example, almost two decades of research addressing transport and deposition processes pointed out the importance of and need to consider reductions in NO_x and VOC emissions, as well as in SO_2. Insights from engineering similarly identified the commercial availability and removal efficiency of alternative technologies for reducing precursor emissions by new and existing sources. Economic analyses and theory informed the various interested parties about the potential costs of various emissions reduction scenarios, as well as prospective benefits if the adverse effects of acid deposition were ameliorated, including the greater economic efficiency of market incentive-based strategies.

This suggests two related questions. First, what was the relative influence of scientific, technical, and economic information in shaping the regulatory program ultimately endorsed by the Congress? Second, did the assessment activities synthesizing the results from the federal government's large-scale, interagency research program conducted under the NAPAP umbrella directly influence the acid deposition provisions in the 1990 CAA amendments? A simple but not necessarily carefully reasoned response asserts that the tremendous amount of information per se and NAPAP specifically were influential in shaping public policy. In fact, however, there are widespread differences of opinion about the overall importance of new information in general and of the NAPAP state-of-science-and-technology reports, or its integrated assessments specifically, for the legislation that ultimately was adopted (NAPAP, 1991a, 1991b).

On balance, a reasonably strong case can be made that new information (some of which was produced under the auspices of NAPAP) defining the scientific, engineering, economic, and institutional dimensions of the acid deposition problem was used in both the agenda-setting and formulation phases of the policymaking processes. Assertions that "science" (i.e., systematic, empirical information) was irrelevant appear groundless. Slightly more than a decade ago, in 1980, few Americans identified acid deposition as an environmental policy problem. Only three years later, a Harris poll found that 63 percent of those questioned were aware of acid rain and approximately 66 percent favored stricter controls on SO_2 emissions. In essence, in the early 1980s, scientific inquiry had transformed the acid deposition issue into a public policy question and played a key role in placing it on the environmental agenda.

Results of Assessment Studies

A number of initial assessment or synthesis studies conducted during the early to mid-1980s concentrated primarily on the state of existing scientific and technical knowledge but also offered some insights into the economics of alternative emissions reduction scenarios. The results were instrumental in identifying regulatory options as well as the nature and extent of ecological effects. As a result, those reports were a major source of current information for decision makers involved in policy development. For instance, when serious deliberations about appropriate policy responses to the acid rain issue began in 1980, there were claims that sulfur emissions were causing an environmental "catastrophe" that was devastating aquatic and terrestrial ecosystems. Counterclaims were being made that acidity was primarily from natural sources, was not causing demonstrable impacts, and acidification levels were not likely to decrease substantially if emissions declined.

Within the first five years of research by NAPAP as well as independent efforts, the policy debate surrounding congressional deliberations recognized the importance of NO_x and VOCs, the localized nature of damage to sensitive surface waters, the highly uncertain role of acidification in forest diebacks, and the increasing marginal costs of control programs, especially for annual emissions reductions greater than 10 million tons.

Although extensive research conducted by the private sector (such as EPRI's ILWAS project and the Living Lakes program) as well as under government auspices was instrumental in characterizing aquatic and terrestrial effects, the results of these studies appear to have had a limited impact on congressional deliberations and the evolving legislation. The continuing controversy surrounding aquatic acidification illustrates this point. While they did not say so explicitly, the 1983 and 1986 National Research Council reports, especially the water chemistry data presented in the 1986 report, fostered the impression that other human activities have much more substantial effects on soil and water chemistry than those producing acid rain, except in a very few high-elevation watersheds that are otherwise undisturbed by humans. The NAPAP National Surface Water Survey conducted during the mid-1980s also yielded late-summer, water chemistry data that reinforced the conclusion that addressing lake acidification ought not be the primary motivation for legislation (Landers et al., 1987; Linthurst et al., 1986). While a paleoacidification study of lakes in the Adirondacks conducted with EPRI funding indicated that most of the lakes with pH values below 6.0 had acidified in the twentieth century, which underscores the importance of man-made sources (Charles et al., 1990), the results of this study, unlike earlier ILWAS results, were not readily available until recently and did not influence the congressional deliberations. As a result, although a tremendous number of studies on effects were conducted and many participants in the policy process knew about the findings, the accumulated evidence about the nature, rate, and magnitude of damage was not a driving force in the congressional deliberations shaping the final legislation.

On the other hand, appraisals of the existing scientific knowledge about atmospheric processes and source-receptor relationships did play a key role. For example, the 1981 National Research Council report was extremely influential in defining emissions reduction goals. The report's recommendation of a 50 percent reduction in H^+ ion deposition to protect sensitive aquatic ecosystems was used to justify proposals for a corresponding 50 percent reduction in SO_2 emissions. The 1983 draft critical assessment review papers requested by the Clean Air Scientific Advisory Committee of EPA's Science Advisory Board (EPA, 1983a), the OSTP peer review panel's report (Nierenberg et al., 1984), and the 1984 OTA report contributed to heightening awareness of NO_x and VOCs in addition to SO_2 as precursor emissions and the transformation of pollutants to acidic compounds. As a result, when they were adopted, the 1990 CAA amendments focused on three pollutants rather than exclusively emphasizing reductions in SO_2 emissions. The 1983 NRC report that concluded that reducing annual emissions across a broad spatial domain would produce corresponding, although not necessarily linearly proportionate, reductions in average annual deposition reinforced a focus on regulatory strategies that limited atmospheric loadings on a yearly basis rather than on a shorter time horizon.

Contribution of Models

Insights from environmental economics also played a major role, if not the dominant one, in shaping the precursor reduction policy ultimately adopted by Congress. A strong case can be made that the congressional debate was dominated by the results of simulation models that projected future emissions of precursors, especially by the electric utility sector, under scenarios with different energy, economic, and regulatory conditions. The Coal and Electric Utilities Model (CEUM) developed by ICF, Inc., provided a fairly detailed representation of the coal and electric utility market and was used extensively by the Environmental Protection Agency, the Department of Energy, the Congressional Budget Office, environmental groups, and industry in their analyses of proposed legislation (ICF, 1989, 1985). Because the results of the model reflected the different assumptions of its users, the findings were easy to compare and the CEUM model gained credibility within policymaking circles. Also, since other models such as the Teknekron

Utility Simulation Model (USM) or NAPAP's Advanced Utility Simulation Model (AUSM) tend to produce roughly comparable results when used to estimate the control costs of emissions reductions, basic conclusions about cost curves were reinforced.

Figure 4 shows that the models agree on the general shape of the cost curve as well as on a dramatic increase in marginal costs (in terms of their magnitude and relative uncertainty) for reductions beyond 10 million tons of SO_2 annually (see Parker, 1985). The 8- to 10-million ton range, with associated costs of $4 billion to $7 billion annually, also coincides with the level under the 1.2 lb/million British thermal unit cap established by NSPS at which eastern coal reserves have to be replaced by western low-sulfur coal and flue-gas desulfurization. Near-term costs are likely to be at the lower end of this range because prices for low-sulfur coal are depressed and not likely to increase dramatically given its oversupply. As a result, economic analysis coupled with interregional political realities suggested an upper bound target for reductions contemplated in the congressional deliberations.

Not surprisingly, reflecting the traditional command-and-control approach

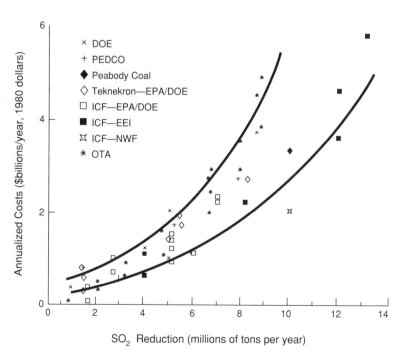

FIGURE 4   Control cost estimates for electric utility emissions reductions. SOURCE: Derived from U.S. Office of Technology Assessment (see also Parker, 1985).

that characterized the U.S. air quality management system throughout the 1970s, the first proposals to emerge during the congressional deliberations over acid deposition controls mandated FGD use to reduce emissions without regard to cost efficiency. A number of analyses, however, suggested that a market incentive-based approach based on emissions trading and marketable permits could achieve comparable reductions in a more economically efficient manner, especially if a phased reduction timetable were used (see Raufer and Feldman, 1987). The emissions inventory developed by NAPAP as part of its Regional Acid Deposition Model (RADM) project indicated that the data were available to implement Title IV of the 1990 CAA amendments, which incorporated the latter approach as the underlying basis for the acid deposition control strategy. In fact, without those data, it is questionable whether interstate, as opposed to intrastate, trading would have been allowed in the final version of the legislation. As a result, economic theory and analysis influenced both the level of reduction mandated and the compliance strategy selected by the Congress.

The examples summarized above illustrate how scientific, technical, and economic information was useful in placing acid deposition on the policy agenda as well as in framing options during the process of formulating the broad guidelines for a regulatory strategy. Information was used to define preliminary options, and the option ultimately chosen represented a modification of the original proposals. These were based on evolving information coupled with White House and congressional willingness to support a regulatory program.

Assessment of NAPAP's Effectiveness

During its 10-year life, NAPAP represented a novel interagency approach to coordinating environmental research in the federal government. It most likely is due some credit for acid deposition policy to the extent that the work to produce useful information was funded under its interagency budget, especially in developing the emissions inventories that form the basis for allocating reductions. It is worth remembering that the original mandate for NAPAP was to produce policy-relevant assessments of the causes and effects of acid deposition as well as to recommend specific policy actions to deal with acid rain. The research being conducted under the NAPAP umbrella was justified as a contribution to a comprehensive program to synthesize understanding of the problem to aid policy choice. NAPAP produced high-quality scientific and engineering studies, particularly in atmospheric sciences and ecosystems research, much of which was at the cutting edge of those sciences.

The NAPAP integrated assessments, however, were not a significant factor in terms of shaping the policy agenda or designing a regulatory strategy. Because it was an executive branch program, NAPAP could only recommend the sitting administration's program rather than independently assess alternative policy choices. This situation frustrated some legislators who had hoped to use NAPAP as an independent source of advice during congressional deliberations. Instead, with few exceptions, the testimony and written documents produced by the NAPAP tended to emphasize scientific and technical information while leaving value judgments and policy implications to others.

In essence, the NAPAP synthesis efforts failed to exert a direct impact on the very policymaking process that provided its formal rationale for existence. However, the widely held expectation that Congress would use NAPAP outputs "directly" is a naive perspective of how scientific, technical, or economic knowledge and policy choice are related. It is incorrect to assume that there is a direct link between good science and good policy (see Regens, 1984). All environmental legislation reflects personal preferences and societal values that shape interpretations of the facts that scientific, technical, and economic information provide. Thus, the political process, not science per se, dictates how much information is enough as well as the conditions under which information guides a given policy decision. When viewed in this light, it is possible to clarify why NAPAP's influence on congressional deliberations was both constrained and largely indirect.

Several reviews of NAPAP concluded that the lack of well-defined information needs coupled with a decentralized management approach tended to make agencies willing to seek funding under NAPAP but precluded the timely collection of valid, reliable data for risk-benefit analyses as a basis for designing a national acid deposition control strategy (EPA, 1983b; NAPAP, 1991c; U.S. General Accounting Office [GAO], 1987). Why then was scientific, technical, and economic information in general influential in shaping the congressional debate while the NAPAP integrated assessments failed to play a central role in policymaking? It is important to answer this question given suggestions that NAPAP might serve as a model for future large-scale federal research efforts to improve understanding of and guide policy development on major environmental concerns, such as global climate change.

Careful appraisal of the NAPAP experience suggests several possible explanations for this outcome. For one thing, policy leaders in the executive branch did not establish at the start the priorities for the areas requiring additional scientific, technical, or economic information to guide policy choices. Moreover, NAPAP lacked true budgetary integration, and much of its research, especially in its early years, amounted to little more than simple relabelling of existing agency programs to fit under the NAPAP umbrella. The joint chairs and NAPAP director had limited control over the design and conduct of the scientific research activities, or the assessments, which were managed by the individual agencies. As a result, in its early years the extent to which the research effort focused on producing information with clearly discernable policymaking value was limited.

Another constraint on the NAPAP's effectiveness was the Reagan administration's use of executive branch personnel associated with NAPAP to support, through claims of scientific uncertainty, its opposition to immediate regulatory interventions. Advocates of intervention asserted that the use of such testimony did little to instill confidence in science as a means of resolving policy choice. This controversy fostered a perception that the NAPAP was little more than a delaying tactic by the Reagan administration to serve a political agenda and that it was under pressure to support political decisions. The acrimony surrounding the executive summary of the 1987 interim assessment did little to assuage such doubts. Unlike the supporting volumes of the 1987 interim assessment report, the executive summary prepared by Dr. J. L. Kulp, at the time NAPAP's scientific director, did not receive any external peer review. Numerous observers asserted that the executive summary offered a highly selective interpretation of the larger body of research findings, and its publication caused NAPAP's assessment efforts to lose considerable credibility. This episode illustrates how individual scientists in key positions can color the debate over the proper interpretation, especially the policy implications, of findings when they advocate specific policies.

The NAPAP's lack of timeliness in producing periodic assessments of the policy relevance of principal scientific, technical, and economic findings was a significant shortcoming. In fact, throughout its existence, the NAPAP encountered major difficulties in meeting its own deadlines for assessment reports. The 1985 preliminary damage assessment was not released because of a change in program leadership and management philosophy. This caused a two-year delay in making the first assessment publicly available, and the discredited 1987 assessment further diminished NAPAP's effectiveness in influencing congressional deliberations. By the time the post-Kulp leadership reestablished the more open decision processes that characterized NAPAP's first few years and enhanced the program's assessment capability, a three-month extension was necessary in order to complete the final 1990 integrated assessment and it was too late for the final assessment to influence legislative outputs. Ironically, the final assessment was released in early 1991, several months after acid rain control legislation was signed into law on November 15, 1990.

On the other hand, the NAPAP's primary scientific and technical conclusions supporting the integrated assessment were presented publicly in February 1990 at an international meeting held in Hilton Head, South Carolina, and the NAPAP had been issuing reports on individual projects throughout its existence. As a result, the major research findings on atmospheric processes, control technology, or effects of acid deposition that might have been useful in designing potential regulatory strategies had gone through an open process of peer review and were readily available to the public and decision makers.

Finally, the speed with which public perception of acid rain as a serious environmental problem continued to increase throughout the 1980s also limited the usefulness of NAPAP's potential contributions to policy decisions. The heightened significance of the acid deposition issue generated pressure for political action and, when the Bush administration proposed acid deposition legislation in 1989, the question of deferring a policy decision until NAPAP's final assessment was released became moot.

LESSONS FOR ENVIRONMENTAL POLICYMAKING

Sensitive environmental issues, by their very nature, create controversy. Once they find a niche in the policymaking process, the existence of scientific, technical, or economic uncertainty may forestall action but ultimately is not likely to preclude regulatory intervention. Careful assessments of existing knowledge that emphasize risk-benefit information, therefore, can map out the advantages and disadvantages of available policy options. This underscores the need to focus on "policy-relevant" research that reduces uncertainty about the probable outcomes of alternative choices. Unless research efforts are guided by the appropriate data requirements for integrated policy assessments, the likelihood of producing useful and timely assessments that inform decision makers during, rather than after, congressional debate is decreased substantially.

Unfortunately, although the NAPAP experience suggests that mission agencies are willing to cooperate in pursuing collaborative research efforts, interagency support for policy assessment activities is more difficult to mobilize and sustain. In large part, this is because the respective agencies have different fundamental mandates that, while allowing cooperation in research, and create strongly divergent views on issues of policy. Despite such obstacles, if the following guidelines are adopted, they are likely to increase the probability that focused research will produce timely, credible, and useful assessments to inform environmental decision making.

• Senior decision makers should identify major policy-relevant questions as a guide before research program planning is started. Follow-up reviews with those decision makers should be used to update key information gaps and uncertainties in policy choices.

• The assessment function should "drive" the research function. Adequate funding and staffing should be provided to integrate assessment activities with research activities. In contrast to the usual pattern of discontinuing research once a policy option has been selected, Congress recognized the need to evaluate policy outcomes by reauthorizing NAPAP under the 1990 CAA in a unique effort to document the benefits of the acid rain control program.

• Recognize that the timeline for policy choice within the political system may differ from the optimal timeline for scientific, technical, or economic inquiry. To be of value to decision makers, information must be available if it is to play a role in policy formulation. Adapt to this by periodically releasing summaries of the state of knowledge and indicate degrees of uncertainty for policy-relevant questions.

• Credibility is critical for policy assessments. To establish and maintain credibility as well as to confront directly the problem of partisan use or misuse of information and its implications, the operational plan and all analytical reports, including executive summaries, should receive external peer review. An outside group of experts should serve as an oversight board, and meetings should be held regularly with the public, academic community, environmental groups, industry, and congressional staff to present scientific findings and assessment results.

Some important lessons in terms of the potential for incorporating new information into environmental policymaking can be learned from the acid deposition experience. They may be useful in developing policies to address other emergent environmental issues, such as global climate change. Those insights are likely to be especially useful if, as some advocate, an interagency approach similar to NAPAP is adopted. Such an approach is appealing, at least in part, because in the case of NAPAP it provided an opportunity for numerous individuals with differing perspectives in the mission agencies as well as in the research community to gain a valuable understanding of the dynamics of the environmental policymaking process.

NOTES

1. To be consistent with popular usage, the term *acid rain* is used in this case study as a catchall for all forms of acidic deposition, unless otherwise indicated.

2. Background interviews for this case study were conducted with individuals representing a wide range of affiliations: the electric utility and coal industries, U.S. Department of Energy, U.S. Environmental Protection Agency, National Acid Precipitation Assessment Program (NAPAP), research scientists, congressional staff, and environmental organizations. I assured those interviewed that they would not be quoted: I feel confident that they were very candid in giving me their opinions.

3. Since acids release hydrogen ions (H^+) in an aqueous solution, the level of acidity typically is measured by the logarithmic pH scale, with pH being equal to the negative $\log_{10}$ of the H^+ ion concentration.

4. By convention, the "natural" acidity value for precipitation is assumed to be pH 5.6; this is a somewhat arbitrary threshold calculated for distilled water in equilibrium with atmospheric carbon dioxide concentrations.

REFERENCES

Beamish, R. J., and H. H. Harvey. 1972. Acidification of the Lochoche Mountain Lakes, Ontario, and Resulting Fish Mortalities. Journal of the Fisheries Research Board of Canada 29:1131–1143.

Bohi, D. R., and D. Burtraw. 1991. Avoiding Regulatory Gridlock in the Acid Rain Program. Journal of Policy Analysis and Management 10:676–684.

Carter, J. E. 1980. Environmental Priorities and Programs - Message to the Congress, August 2, 1979. Public Papers of the Presidents of the United States: Jimmy Carter 1979 Book II. Washington: Government Printing Office.

Charles, D. F., M. W. Binford, E. T. Furlong, R. A. Hites, M. J. Mitchell, S. A. Norton, F. Oldfield, M. J. Patterson, J. P. Smol, A. J. Uutala, J. R. White, D. R. Whitehead, and R. J. Wise. 1990. Paleolimnological Investigations of Recent Acidification of Lakes in the Adirondack Mountains, New York. Paleoecology 3:195–241.

Cogbill, C. V., and G. E. Likens. 1974. Acid Precipitation in the Northeastern United States. Water Resources Research 10:1133–1137.

Cowling, E. B. 1982. Acid Precipitation in Historical Perspective. Environmental Science & Technology 16:110A-123A.

ICF. 1985. Analysis of Sulfur Dioxide and Nitrogen Oxide Emission Reduction Alternatives with Electric Rate Subsidies, Report prepared for the National Wildlife Federation. Washington, D.C.: ICF, Inc.

ICF. 1989. Economic Analysis of Title V (Acid Rain Provisions) of the Administration's Proposed Clean Air Act Amendments (H.R. 3030/S. 1490), Report prepared for the U.S. Environmental Protection Agency. Washington, D.C.: ICF, Inc.

Landers, D. H., J. M. Eilers, D. F. Brakke, W. S. Overton, P. E. Kellar, M. E. Silverstein, R. D. Schonbrod, R. E. Crowe, R. A. Linthurst, J. M. Omernik, S. A. Teague, and E. P. Meier. 1987. Characteristics of Lakes in the Western United States, Vol 1. EPA/600/3-86/054a Washington, D.C.: U.S. Environmental Protection Agency.

Linthurst, R. A., D. H. Landers, J. M. Eilers, P. E. Kellar, D. F. Brakke, W. S. Overton, E. P. Meier, and R. E. Crowe. 1986. Characteristics of Lakes in the Eastern United States, Vol 1. EPA/600/4-86/007a. Washington: U.S. Environmental Protection Agency.

National Acid Precipitation Assessment Program. 1985. Annual Report, 1985. Washington, D.C.: National Acid Precipitation Assessment Program.

National Acid Precipitation Assessment Program. 1991a. Acidic Deposition: State of Science and Technology. Washington, D.C.: Government Printing Office.

National Acid Precipitation Assessment Program. 1991b. 1990 Integrated Assessment Report. Washington, D.C.: Government Printing Office.

National Acid Precipitation Assessment Program. 1991c. The Experience and Legacy of NAPAP: Report of the Oversight Review Board. Washington: National Acid Precipitation Program.

National Research Council. 1981. Atmosphere-Biosphere Interactions: Toward a Better Understanding of the Ecological Consequences of Fossil Fuel Combustion. Washington, D.C.: National Academy Press.

National Research Council. 1983. Acid Deposition: Atmospheric Processes in Eastern North America: A Review of Current Scientific Understanding. Washington, D.C.: National Academy Press.

National Research Council. 1986. Acid Deposition: Long-Term Trends. Washington, D.C.: National Academy Press.

National Research Council of Canada. 1981. Acidification in the Canadian Aquatic Environment: Scientific Criteria for Assessing the Effects of Acidic Deposition on Aquatic Ecosystems. Ottawa: National Research Council of Canada.

Nierenberg, W. A., W. C. Ackerman, D. H. Evans, G. E. Likens, R. Patrick, K. A. Rahn, F. S.

Rowland, M. A. Ruderman and S. F. Singer. 1984. Report of the Acid Rain Peer Review Panel. Washington, D.C.: Office of Science and Technology Policy.

Oden, S. 1968. The Acidification of Air and Precipitation and Its Consequences in the Natural Environment. Ecology Committee Bulletin no. 1, Swedish National Science Research Council.

Organization for Economic Cooperation and Development. 1977. The OECD Programme on Long-Range Transport of Air Pollutants. Paris: OECD.

Overrein, L. N., H. M. Seip, and A. Tollan. 1980. Acid Precipitation-Effects on Forest and Fish. Final Report of the SNSF Project, 1972–1980, SNSF Research Report no. 19. Oslo-As: SNSF Project.

Parker, L. B. 1985. Response to Hoff Stoffer. Pp. 49–58 in Acid Rain, P. Mandelbaum, ed. New York: Plenum Press.

Raufer, R. K., and S. L. Feldman. 1987. Acid Rain and Emissions Trading. Totowa, N.J.: Rowman & Littlefield.

Regens, J. L. 1984. Acid Rain: Does Science Dictate Policy or Policy Dictate Science? Pp. 5–19 in T. D. Crocker, ed. Economic Perspectives on Acid Deposition Control. Woburn, Mass.: Butterworth Scientific Publishers.

Regens, J. L. 1989. Congressional Cosponsorship of Acid Rain Controls. Social Science Quarterly 70:505–512.

Regens, J. L., and R. W. Rycroft. 1986. Options for Financing Acid Rain Controls. Natural Resources Journal 26:519–549.

Regens, J. L., and R. W. Rycroft. 1988. The Acid Rain Controversy. Pittsburgh: University of Pittsburgh Press.

Smith, R. A. 1872. Air and Rain: The Beginnings of Chemical Climatology. London: Longmans, Green.

Swedish Ministry of Foreign Affairs and Swedish Ministry of Agriculture. 1972. Pollution Across National Boundaries: The Impact on the Environment of Sulfur in Air and Precipitation. Sweden's Case Study for the United Nations Conference on the Human Environment. Stockholm: Royal Ministry of Foreign Affairs & Royal Ministry of Agriculture.

U.S.-Canada Work Group 1. 1982. Impact Assessment: Final Report. Washington, D.C.: U.S. Environmental Protection Agency.

U.S.-Canada Work Group 2. 1982. Atmospheric Science and Analysis: Final Report. Washington, D.C.: U.S. Environmental Protection Agency.

U.S.-Canada Work Group 3B. 1982. Emissions, Costs, and Engineering: Final Report. Washington, D.C.: U.S. Environmental Protection Agency.

U.S. Congress Office of Technology Assessment. 1984. Acid Rain and Transported Air Pollutants: Implications for Public Policy. Washington, D.C.: Government Printing Office.

U.S. Department of State. 1981. Transboundary Air Pollution: Memorandum of Intent between the United States of America and Canada, signed at Washington, August 5, 1980. Treaties and Other International Acts Series 9856. Washington, D.C.: U.S. Department of State.

U.S. Environmental Protection Agency. 1983a. The Acid Deposition Phenomenon and Its Effects: Draft Critical Assessment Review Papers. Washington, D.C.: U.S. Environmental Protection Agency.

U.S. Environmental Protection Agency. 1983b. Report of the Ad Hoc Committee to Review the National Acid Precipitation Assessment Program (NAPAP). Prepared for the Science Advisory Board. Washington, D.C.: U.S. Environmental Protection Agency.

U.S. General Accounting Office. 1987. Acid Rain Delays and Management Changes in the Federal Research Program. GAO/RCED-87-89. Washington, D.C.: Government Printing Office.

Keeping Pace with Science and Engineering. 1993.
Pp. 189–220. Washington, DC: National Academy Press.

Formaldehyde Science: From the Laboratory to the Regulatory Arena

Susan W. Putnam and John D. Graham

In 1979 the Chemical Industry Institute of Toxicology (CIIT), a research organization funded primarily by a consortium of chemical corporations, released the interim results of rodent bioassays indicating that exposure to formaldehyde (HCHO) causes nasal cancer in rats. This implication of the chemical as an animal carcinogen raised immediate questions about its potential as a human carcinogen. Over the next decade, several regulatory agencies struggled with the problem of what to make of these data, and, ultimately, what to do about formaldehyde.

While not immediately accepted in all circles, CIIT's initial bioassay data on HCHO ultimately became the foundation of risk assessments used by all of the relevant regulatory agencies. The data left many questions unanswered, however, especially the mechanisms of carcinogenicity and their implications for assessing human risk. In an effort to answer these questions, CIIT expanded its formaldehyde research program into pharmacokinetics and mechanism of action. These new studies explored biologic issues not only in the rat but also in primates, a species more similar to humans in physiology and metabolism. Simultaneously, a large epidemiologic study provided new insights into HCHO exposure and human cancer.

The pharmacokinetic and mechanistic data have not been readily accepted in the regulatory arena. There also has been controversy over the interpretation of the latest epidemiologic information. The question of how to interpret all these data in risk assessment has been a hot regulatory issue. Should the new methodologies and results be considered established science, acceptable and appropriate for regulatory decisions?

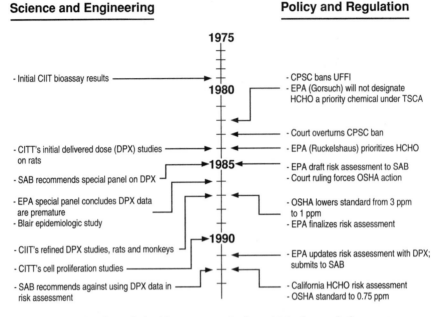

FIGURE 1   Timeline of significant events in formaldehyde regulation.

The way in which one interprets the new science can make a big differ-ence in human risk estimates. If the new animal results are incorporated into risk assessments for formaldehyde using one of the most prevalent interpretations of the data, the human risk estimates for the chemical are 10 to 100 times lower than if they are not included. Advocates of the animal data champion the scientific advances that they embody; opponents fear the data underrate the risk that formaldehyde poses for humans.

This case study examines the efforts (and reluctance) to incorporate the biologic data into the regulatory arena. Tracing the scientific and regulatory histories of formaldehyde, the study explores the key issues and current debate concerning the appropriateness of the new science for use in regula-tory action. Figure 1 shows the timeline of significant events in the formal-dehyde story.

BACKGROUND—A UBIQUITOUS AND IRRITATING CHEMICAL

Formaldehyde is an omnipresent and versatile chemical that is pro-duced naturally by the human body (Cascieri and Clary, 1992). It is manu-factured and used by industry to create a variety of products, such as plastics, adhesive resins for plywood and particleboard, permanent-press fabrics, and numerous household products. It is also widely used as an embalming fluid. In addition, HCHO is generated by multiple sources, such as incomplete combustion, and the chemical is a major component of cigarette smoke. As a consequence, formaldehyde is ubiquitous in the ambient air of both indoor and outdoor environments, often in significant concentrations.

The irritant properties of formaldehyde have long been recognized. Consumers exposed to products containing the chemical, as well as workers employed in industries using the substance, may be susceptible to irritation of the eyes, skin, or respiratory system as a result of exposure to the chemical. Because of these irritant effects, much research had been focused on the acute toxicity of formaldehyde (National Research Council [NRC], 1980).

The acute effects of the chemical have been fully acknowledged by the regulatory agencies. Based on irritant effects, the Occupational Safety and Health Administration (OSHA) adopted, in 1972, a worker exposure standard for the chemical of 3 parts per million (ppm) (29 C.F.R. 1910.1000[b] Table Z-2). The Consumer Product Safety Commission (CPSC) was similarly troubled by the irritant effects that consumers were experiencing in their homes. Of particular concern was consumer exposure to urea-formaldehyde foam insulation that was being installed in many homes by the late 1970s to conserve energy. These were often mobile or factory-built homes for elderly or low income residents.

THE INDICTMENT—IS FORMALDEHYDE A HUMAN CARCINOGEN?

Although the irritant effects of formaldehyde were well established, few studies had explored the question of the chemical's carcinogenicity. The CIIT rodent bioassay was the first persuasive indication of formaldehyde's carcinogenic potential. The release of the interim bioassay results in 1979 unleashed a new and volatile chapter in the scientific history of formaldehyde. The results spawned a wealth of both scientific and regulatory debate.

The CIIT study was a two-year inhalation bioassay exposing both rats and mice to several exposure levels of HCHO (Kerns et al., 1983). The bioassay results are reported in Table 1. By the end of the study, half of the 206 rats exposed to the highest concentration of the chemical, 14.3 ppm, had developed squamous cell carcinomas of the nasal cavity. Two rats developed similar tumors at the next lower exposure level, 5.6 ppm. At the lowest exposure levels, no malignant tumors were apparent. Neither were there many tumors in the mice, even at the 5.6 and 14.3 ppm concentrations. There was an elevated incidence of polypoid adenomas (benign tumors) in the experimental rats at all the exposure levels, although no dose-response relationship was apparent.

TABLE 1   Cancer Incidence in Rodents Following Inhalation
of Formaldehyde

Formaldehyde	Number of Tumors/Animals at Risk[b]	
Concentration (ppm)[a]	Rats	Mice
0	0/208 (0%)	0/72 (0%)
2	0/210 (0%)	0/64 (0%)
6	2/210 (1%)	0/73 (0%)
15	103/206 (50%)	2/60 (3.3%)

NOTE: Inhalation was for 6 hours/day, 5 days/week. The study was initiated
with 960 Fischer-344 rats and 960 B6C3F1 mice, evenly divided by sex into
treatment groups. Duration of the study was 24 months.

[a]Target concentrations.   Actual average measured concentrations were 0,
2.0, 5.6, and 14.3 ppm.
[b]Actual number of animals exposed to formaldehyde up to, and including,
the interval when the first squamous cell carcinomas were found (11–12 months
for rats; 23–24 months for mice).

SOURCE: Gibson (1983, p. 297).

The results of the long-term bioassay experiments were interpreted to
demonstrate that, at least for Fischer-344 rats (a specific strain) exposed to
high concentrations of the chemical, formaldehyde was an animal carcinogen. This conclusion was later corroborated by other studies and expanded
to include the induction of similar squamous cell tumors in different strains
of rats (Albert et al., 1982; Tobe et al., 1985).

The CIIT data raised many questions. Of particular concern was the
nonlinear tumor response as the HCHO concentrations were increased. The
highest exposure level, 14.3 ppm, was two and a half times higher than the
next level (5.6 ppm), yet there was a fiftyfold difference in tumor response.
To complicate matters further, there were no malignant tumors at the lowest
exposure levels. What did this imply for the shape of the dose-response
curve, particularly at low concentrations? Did this suggest that a threshold
exists for the chemical's carcinogenic activity? Was formaldehyde "safe" at
low exposure levels?

Another concern was raised about the paucity of tumors in the mice.
Were the squamous cell carcinomas specific to the physiology and metabolism of the rat (e.g., the rat is an obligatory nose breather), or might other
species have similar tumor responses? With the rat and mouse tumor rates
differing so greatly, how should the results be extended to other species?
Most important, given such issues, how should these data be interpreted for
humans, who were most generally exposed to chronic low concentrations of

HCHO, although levels in the workplace and home were often within a factor of ten of 5.6 ppm?

NEW REVELATIONS

In an effort to answer these and other questions, major scientific investigations on formaldehyde continued in several critical areas. Some of this research, such as the studies exploring pharmacokinetics and mechanisms of cancer, offered groundbreaking methodologies and ideas that were at the frontiers of science. Other work, such as the several major epidemiologic studies, contributed much larger data sets and analytical power than were previously available. Taken together, the studies offered significant new insights on the potential human carcinogenicity of formaldehyde.

Delivered Dose (DPX)

One of the major areas of formaldehyde research at CIIT was pharmacokinetics. Using biochemical methods, scientists can trace the uptake, metabolism, and distribution of HCHO in the body, including the interaction of the chemical with DNA in the target cells of various tissues.

The first pharmacokinetic studies, performed by Casanova, Heck, and their colleagues at CIIT, explored the relationship between the amount of formaldehyde that the animals were exposed to—the administered dose— and the dose of the chemical that actually reached the target tissue cells in the rat nose—the delivered dose. Using radiolabeled HCHO, the researchers measured the amount of covalent binding of the chemical to rat nasal mucosal DNA and formation of DNA-protein cross-links[1] (Casanova-Schmitz et al., 1984). It was hypothesized that the covalent binding would not only be a surrogate measure of the delivered concentration of the chemical, but that it was also related in some way to the observed nasal tumors in the original rodent studies. The covalent binding was assessed for acute (6-hour) exposures in the Fischer-344 rat. The concentrations of administered formaldehyde were similar to those used in the initial rodent bioassay.

The results of these studies indicated that the delivered dose was not linearly related to the administered concentration of the chemical. This nonlinearity was particularly manifest for the lower exposures, as Figure 2 indicates. For example, the concentration of formaldehyde covalently bound to DNA at 6 ppm was 10.5-fold higher than at the 2 ppm exposure level. In other words, the DNA binding at the 2 ppm level was significantly lower than the amount predicted by drawing a straight line from 6 ppm to the origin.

These data were less than ideal for use in risk assessment. It is difficult to extrapolate from short-term exposures in small groups of animals to the

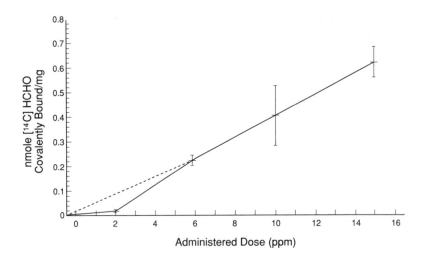

FIGURE 2   Delivered dose results.   SOURCE: Casanova-Schmitz et al. (1984, p. 38).

chronic exposures generally experienced by humans. However, the indication of nonlinearity in the DNA-protein cross-link (DPX) data held significant implications for the cancer risk associated with formaldehyde. If these data were an accurate representation of the dose reaching the tissue after exposure to low concentrations of the chemical, then there may actually be less response at low doses than the linear curve predicted. This idea was in fact illustrated in a risk assessment performed by Starr and Buck of CIIT. The analysis, summarized in Table 2, showed that cancer risk estimates using the new DPX data were at least an order of magnitude lower than those using the administered doses (Starr and Buck, 1984).

Subsequent studies on glutathione depletion explored the effect of glutathione as a defense mechanism in the formation of DNA-protein cross-links (Casanova and Heck, 1987). Other work examined the "isotope effect" in the use of formaldehyde labeling to quantify the DNA cross-linking process (Heck and Casanova, 1987), and augmented the measurement of this phenomenon by the use of improved experimental procedures and technology (Casanova et al., 1989). This new work built upon the initial concept explored in the earlier delivered-dose studies and developed superior measurement schemes to quantify the results. From CIIT's perspective, these additional studies served to further support the nonlinear relationship between the administered and delivered doses of the chemical. The CIIT scientists argued that this relationship may be even more nonlinear than was initially perceived.

In addition to the pharmacokinetic experiments in rats, CIIT researchers conducted similar experiments on DNA-protein cross-links in six rhesus monkeys (Casanova et al., 1991). Since there are marked differences between humans and rats with respect to nasal anatomy and respiratory physiology, monkeys were selected as experimental subjects in an effort to use a species more closely related to humans. The rhesus monkey is biochemically and physiologically similar to humans, particularly in the important target site areas for formaldehyde (e.g., the nose and upper respiratory tract).

The results of the monkey studies were quite similar to what was predicted using the rat studies and interspecies scaling factors (e.g., body weight). There were important differences in the results, however. The monkey data indicated a lower rate of DNA-protein cross-link formation than in the rat, particularly for the low concentrations of formaldehyde. Monkeys also exhibited a wider distribution of cross-links and lesions in the upper respiratory tract than did the rats (Casanova et al., 1991).

The monkey DPX data also held significant implications for HCHO cancer risks. The lower rate of cross-link formation and larger target area in the monkey were used to predict a lower cancer risk for primates, and, ultimately, an even lower one for humans than was estimated using the rat data. For example, in the 1987 EPA risk assessment for formaldehyde that did not incorporate the pharmacokinetic data, the upper-bound risk of human cancer at 0.1 ppm exposure was 1.6×10^{-2}. In the 1991 analysis, the upper-bound risk estimate at the same exposure level using the rat DPX data was 2.8×10^{-3}. Using the monkey-based data, the risk estimate was 3.3×10^{-4}. The DPX data, especially those for the monkey, result in a much smaller estimate of human risk than do the initial bioassay data: a sixfold

TABLE 2   Formaldehyde Risk Assessment[a]

Exposure Level (ppm)	Dose Metric	Maximum Likelihood Estimate of Risk	Upper 95% Confidence Limit
0.1	Administered	2.5×10^{-7}	1.6×10^{-4}
	Delivered	4.7×10^{-9}	6.2×10^{-5}
1.0	Administered	2.5×10^{-4}	1.8×10^{-3}
	Delivered	4.7×10^{-6}	6.2×10^{-4}

[a]Three-stage multistage model fitted to CIIT rat bioassay data using either administered or delivered dose as the dose metric. Risk is the lifetime probability of a malignant tumor for chronic exposure at the stated level.

SOURCE: Starr and Buck (1984).

decrease with the rat data, a fiftyfold decrease with the monkey data (EPA, 1991). Other models using the pharmacokinetic data on HCHO consistently produce similar or even lower estimates (Starr, 1990).

Mechanisms of Carcinogenesis

While pharmacokinetic studies are useful, the key area of scientific investigation on formaldehyde involved the mechanisms of carcinogenesis. What are the biological mechanisms responsible for the growth of the nasal carcinomas in rats? At the heart of the research lay two viable hypotheses: mutagenesis and cell proliferation.

Scientists ascribing to the mutagenicity theory argue that contact between formaldehyde and the DNA in cells could induce mutations in the critical genes that could ultimately result in the growth of cancerous tumors. There have been several studies exhibiting the mutagenic potential of the chemical (Goldmacher and Thilly, 1983). Proponents of this work suggested that the nasal carcinomas seen in the rat bioassays may be the result of the mutagenic capability of formaldehyde. However, some scientists saw formaldehyde as only a weak mutagen (Consensus Workshop on Formaldehyde, 1984). Because the chemical is so highly reactive in the body, it is difficult to accurately explore its mutagenic potency.

Mutagenicity is important in assessing human risk. If one believes in the mutagenic capability of formaldehyde to the extreme, then every molecule of the chemical has the potential to induce mutations in the DNA with which it comes in contact. While cancer development is a multistage process, there may be other processes going on in the body to which the chemical exposure adds the final step. Assuming repair mechanisms are imperfect, even low exposures to formaldehyde could induce cancerous tumors. No concentration would then be safe for human exposure. Proponents of this view argue that there may be a linear dose-response curve; if so, there would be some potential risk at even the smallest exposure levels. Although there were no tumors seen at the low doses in the animal bioassays, the potential for tumors occurring in a larger sample of rats tested at these concentrations could not be ruled out.

Another theory of formaldehyde carcinogenesis focuses on cell proliferation. Proponents argue that with the increased cell growth and division induced by the onslaught of a toxic substance, there is an increased frequency of spontaneous DNA mutations. Also, because of the high rate of cell reproduction, there is less time for DNA defense mechanisms to repair the critical mutations that could lead to a carcinogenic response. Elevated rates of cell proliferation may enhance both the likelihood of interaction of formaldehyde with DNA and the fixation of adducts before DNA repair could occur (Monticello et al., 1989). Since cell proliferation is often considered a nonlinear process, the implications for risk assessment are profound. Several studies in the late 1980s and early 1990s (e.g., Monticello and Morgan, 1990; Monticello et al., 1991; Swenberg, 1986) explored this phenomenon. Researchers at CIIT examined the relationship between the rate of formaldehyde-induced cell proliferation and the cancerous tumors seen in the initial bioassays. The studies demonstrated that acute and subchronic exposure to formaldehyde induced nasal epithelial lesions and increases in surface cell proliferation rates in rat nasal passages at HCHO concentrations of 6 ppm and higher. As Figure 3 indicates, significant elevations in cell proliferation rates were not detected at concentrations lower than 6 ppm. This result correlated well with the lack of tumors observed below the 5.6 ppm concentration in the earlier bioassay. Increased cell proliferation rates beginning at 6 ppm were also seen in monkeys (Monticello et al., 1989).

CIIT scientists interpreted the correlation between tumor responses in the original bioassay and sustained increases in cell proliferation as evidence that cell proliferation plays a causative role in the carcinogenic process. If this were the case, then the absence of malignant tumors below the 6 ppm exposure level may be related to the lack of detectable increases in cell proliferation rates. The implication of this hypothesis is that there may

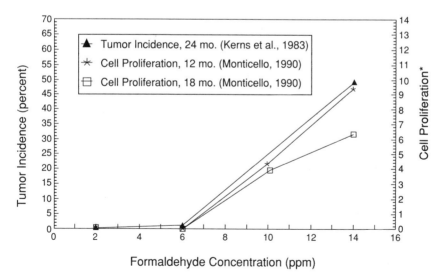

FIGURE 3 Tumor incidence and cell proliferation in rats exposed to formaldehyde. *Cell proliferation is measured by the mean unit length labeling index at nasal level II (fold increase over control). SOURCE: Environmental Protection Agency (1991, p. 30).

be some level—a threshold level—below which no stimulation or increased cell proliferation, and hence no cancer, occurs. Like the DPX data, these new data suggested a nonlinear dose-response curve for formaldehyde at low concentrations. In contrast to those believing in the mutagenic potential of the chemical, proponents of cell proliferation argued that exposure to formaldehyde may be safer at low concentrations than is predicted by linear models.

Cell proliferation data have as yet been difficult to incorporate into risk assessment procedures. Critics assert that while tumor formation as a result of HCHO exposure is preceded by increased cell proliferation, increased cell proliferation may not necessarily cause tumor formation. Also, the current risk assessment process uses mathematical models based on the principles of mutagenesis. The models incorporate a linear dose-response curve, which projects some level of risk even at very low concentrations of the chemical. Preliminary speculation, however, theorizes that if the cell proliferation data with their potential threshold level were incorporated into the analysis, the risk at low concentrations would be reduced, possibly to zero. The risk estimates for human cancer would be lower. Using both the cell proliferation and delivered-dose data, the overall implication for human risk assessment is that inhalation of small doses of formaldehyde may not be as harmful as risk assessors at regulatory agencies had originally predicted.

Another possibility is that cell proliferation exacerbates the incidence of tumors at high doses even though mutagenicity plays some role at low doses. Mutagenesis and cell proliferation are not mutually exclusive ideas. The hypothesis that both these factors play a role is consistent with linearity at low doses and curvature at high doses. If both mechanisms are operating, the unknown slope of the dose-response curve at low doses is critical to risk assessment.

Epidemiology

While a variety of studies were being conducted on laboratory animals, there were several new studies investigating HCHO's effects on humans as well. These epidemiologic studies offered large data sets with which to explore the carcinogenic potential of the chemical, including any nasal tumors similar to those seen in rats.

Many epidemiologic studies have been conducted on formaldehyde over the past several decades. Using both cohort and case-control designs, the studies focused primarily on groups of workers in a variety of occupational settings. The results of these studies were interpreted as suggesting that HCHO may be a human carcinogen for certain groups of professionals (Consensus Workshop on Formaldehyde, 1984). Based on these studies—in conjunction with the animal data—the International Agency for Research on Cancer (IARC) and the U.S. Environmental Protection Agency (EPA) concluded that limited evidence existed for an association between formaldehyde and human cancer, and classified the chemical as a "probable" human carcinogen.

As with many epidemiologic studies, however, the human studies of formaldehyde were criticized as suffering from certain limitations. The major drawbacks were perceived to include small sample sizes, insufficient follow-up of the exposed populations, a low statistical power to detect small relative risks for rare cancers, and an inability to separate the effects of formaldehyde from the myriad confounding substances to which the subject populations were simultaneously being exposed—especially cigarette smoke and particulates (EPA, 1987). Despite these problems, the epidemiologic data base for formaldehyde was much larger than many up to that date and suggested directions for further study.

In an effort to provide more conclusive epidemiologic data on formaldehyde, several large studies were conducted on the chemical in the mid-1980s (Blair et al., 1986; Stayner et al., 1988; Vaughan et al., 1986). These studies were designed specifically to detect moderate elevations in underlying cancer risk among populations with HCHO exposure in the workplace. According to EPA's interpretation, all three studies revealed statistically significant elevations in the risk of site-specific upper respiratory cancers with measures of HCHO exposure (EPA, 1991). These studies were not without their critics, however, particularly for their failure to control for cigarette smoking.

The most notable and widely cited of these studies was that conducted by Blair and colleagues at the National Cancer Institute, DuPont, and Monsanto (Blair et al., 1986). The Blair cohort study followed more than 26,500 workers (approximately 600,000 person-years of data) in 10 plants that either produced or used formaldehyde. The HCHO exposure level of the workers varied, probably averaging about 0.5 ppm for the exposed members of the cohort as a whole. The study did not control for smoking.

As shown in Table 3, the results of the study indicated slight excesses of cancer in the upper respiratory tract and lungs for persons occupationally exposed to formaldehyde. The results also suggested that the risk of cancer in the nasopharyngeal and sinonasal cavities may be enhanced with simultaneous exposure to particulates or wood dust. The authors determined, however, that the tumors did not show a consistently rising incidence with level of exposure. They interpreted the results of the study as providing little evidence that cancer mortality was associated with formaldehyde exposure at levels experienced by the workers in the plants under focus.

The new study was not without controversy, however. Some members of the study's external review panel, as well as labor unions and other

TABLE 3   Mortality from Selected Cancers for White Men

Cancer	Observed	Expected	SMR[a]
Respiratory System			
Exposed	215	192	112
Nonexposed	53	56	95
Buccal Cavity and Pharynx			
Exposed	18	19	96
Nonexposed	3	6	54
Larynx			
Exposed	12	8	142
Nonexposed	4	3	[b]
Lung			
Exposed	201	182	111
Nonexposed	49	53	93

NOTE: The number of persons involved was 16,962 exposed, 5,562 nonexposed. Number of person-years of exposure: 378,497 exposed, 100,237 nonexposed.

[a]Standardized mortality ratio (SMR) is the number of observed cancers divided by the number of expected cancers. An SMR of 100 is interpreted as the norm; the observed number of cancers is the same as the expected number.

Given the occupational cohort of the study, however, there may be some bias due to the "healthy worker effect," whereby the study population of workers may be healthier than the average population overall. If this phenomenon were present, then fewer cancers would be expected for the healthy study population. This would in turn raise the standard mortality ratios for the study results.

[b]SMR is not presented when the observed and expected numbers are less than 5.

SOURCE: Blair et al. (1986, from Table 2, pp. 1075–1077).

groups, criticized the negative interpretation of the results and accused the authors of collaborating too closely with industry. The lack of a consistent dose-response relationship was not seen as detracting from the positive findings presented by the study. Critics of Blair's interpretation also argued that the statistically significant excess of lung cancer in workers (who had an exposure latency of at least 20 years) contradicted the authors' characterization of little or no evidence of carcinogenesis (Jasanoff, 1990).

Blair and colleagues subsequently further evaluated some of the data. They stated, for example, that even though the numbers were small, the pattern for nasopharyngeal cancer suggested that simultaneous exposure to formaldehyde and particulates may be a risk factor (Blair et al., 1987). This conclusion was met with opposition as well, however, this time from the owners of the plant where most of the tumors occurred (Collins et al., 1987, 1988).

Despite the controversies that ensued over its interpretation, the Blair study is perceived to be an important piece of epidemiologic research. Considering the study's size, cost, and methodological care, it is doubtful that a similar study of such proportions will soon be conducted on formaldehyde. The data set yielded by the study is enormous and is still being evaluated (e.g., Blair et al., 1990).

The new epidemiologic data hold important implications for risk assessment as well. Although the data are difficult to quantify for risk assessment estimates, some analyses have been done. The EPA, for example, developed some rough comparisons (using the Blair data) suggesting that observed human risks were concordant with animal-based risk estimates using administered dose levels. The cancer risk estimates using human data were, however, significantly higher than those obtained using the DPX models (EPA, 1990).

THE CONSEQUENCES—THE AGENCIES RESPOND

To relate the story of regulatory action on HCHO to the new scientific information, it is necessary to include a brief synopsis of agency reactions to the initial bioassay data. Several regulatory agencies have jurisdiction over the chemical, most notably the Consumer Product Safety Commission, the Occupational Safety and Health Administration, and the Environmental Protection Agency. Each of the agencies faced the same scientific data sets, yet their responses were curiously unique.

While agencies were aware of the irritant effects of the chemical, it was the release of the CIIT bioassay data in 1979 that brought formaldehyde to the regulatory forefront. The demonstration that the chemical was an animal carcinogen injected a new level of both urgency and controversy into the regulatory debate. Cancer policy guidelines that had been adopted in the late 1970s stated that, in the absence of adequate human data, regulators were to interpret positive and valid animal bioassay data as providing presumptive evidence of human carcinogenic risk (U.S. Regulatory Council, 1979). Given the high level of economic investment tied to such a widely used chemical, there was a natural hesitancy in the agencies to regulate. The new cancer guidelines, however, required that the agencies take a new look at the chemical and its potential risk to humans.

Although many of the agency deliberations and decisions concerning formaldehyde were concurrent and highly intertwined, the events of each agency will be treated separately here in an effort to make the story less confusing. Also, while there were some interagency efforts to coordinate their approach to HCHO risk assessment, such as the Interagency Regulatory Management Council Workgroup on Formaldehyde, many of the resultant decisions appeared to be quite unique to specific agencies. Finally, HCHO has had a long and involved history at many of the agencies. For the sake of brevity, this study does not attempt to cover the complete story at each agency. Instead, there is an effort to cover the regulatory history most related to the new scientific information at issue. The EPA receives the greatest focus here because that was where much of the controversy over the new studies ensued.

Consumer Product Safety Commission

After the release of the rodent bioassay data, the Consumer Product Safety Commission took the most immediate and ambitious regulatory action. Buoyed by the conclusion that the chemical should be assumed to pose a carcinogenic risk to humans (Federal Panel on Formaldehyde, 1982), the commission banned the use of urea-formaldehyde foam insulation (UFFI) in all homes and schools in 1982 (CPSC, 1982). The decision was supported by a quantitative assessment indicating a risk of up to 1.8 additional cases of human cancer for every 10,000 homes using urea-formaldehyde foam insulation. The ban was subsequently overturned the following year by the Fifth Circuit Court of Appeals, on the court's determination that the CPSC had failed to support its action with substantial evidence. The court decision stated that the CPSC had erred in relying solely on the CIIT study to establish the risk of human carcinogenicity (*Gulf South Insulation* v. *CPSC*, 1983). Nevertheless, the early efforts of the CPSC represent an ambitious and prompt response to the demonstration of the chemical's carcinogenicity in animals.

The CPSC's response to the initial CIIT pharmacokinetic model and data on formaldehyde, on the other hand, was less than enthusiastic. The commission's chief scientist responsible for formaldehyde, Murray Cohn, was one of the most vocal opponents to the use of the initial Casanova studies in the risk assessment process. As the lead author of a published critique of the CIIT studies (Cohn et al., 1985), Cohn was concerned about both the relevance of the short-term experiments to the chronic exposures experienced by humans and the relationship of the DNA-protein cross-links to carcinogenesis. Also, he perceived that there was too much "background noise" in the studies to sort out what was really going on. His conclusion was that the methodologies used were weak and that the data were too preliminary to include in the risk assessment process (personal communication with Cohn, 1992).

With the release of the more recent pharmacokinetic studies, the outlook of the commission—or at least of its key scientist—has changed. Today Cohn argues that the new studies incorporate improved procedures that have helped to clear up some of the methodological issues that CPSC was concerned about in the original Casanova research. The commission has not performed an official risk assessment using a pharmacokinetic model and the new data, but it has done unofficial, back-of-the-envelope analyses that do incorporate them. Were a new HCHO regulatory issue to arise, the new analyses would be ready to go (personal communication with Cohn, 1992).

It remains to be seen whether formaldehyde will resurface as a key issue for the CPSC. Urea-formaldehyde foam insulation has been largely phased out of the market, as well as removed from many of the homes in which it was used. The commission is currently working with industry on a voluntary emissions standard for urea-formaldehyde particleboard, but this has not required that a formal new risk assessment be done.

Occupational Safety and Health Administration

In contrast to the CPSC's prompt action following the release of the bioassay results, the Occupational Safety and Health Administration was slower to respond to the CIIT data. Leaders at the agency argued that the cancerous tumors in the bioassay results appeared at a higher exposure level than that faced by humans in the workplace (Graham et al., 1988). It was not until a federal district court ordered OSHA to formally consider initiating a proceeding for a new permanent rule on formaldehyde—in response to a suit against the agency brought by the United Auto Workers (UAW) and other groups seeking an emergency temporary standard for HCHO—that the agency took any formal action on the chemical. The proposed rulemaking procedure for HCHO that ensued in 1985 was based entirely on the chemical's irritant effects, however, not on its cancer risk (OSHA, 1985).

It was not until 1987 that the agency used the CIIT bioassay data in regulatory proceedings, when it lowered its standard for formaldehyde from 3 ppm to 1 ppm as an 8-hour, time-weighted average (OSHA, 1987b). The agency's quantitative risk assessment incorporated the CIIT data and indicated that by lowering the standard from 3 to 1 ppm, the lifetime risk of cancer per 100,000 workers would decrease in the range of 70 to 570 cases.

The reduction in cancer risk was not inexpensive. OSHA estimated that the annual cost of the new standard would be approximately $62 million. This cost was judged to be economically feasible, however, based on an industry-specific analysis of revenue and profit ratios. The new standard was also perceived to be technologically feasible. The agency foresaw that many of the industries could pass on some of the costs of compliance to purchasers of their products, and the new standard would not adversely affect the viability of most of the firms in the relevant industries. Also, the new standard was not perceived to have a significant adverse effect on a substantial number of small entities nor on the economy as a whole (OSHA, 1987c).

While the 1987 rulemaking process for formaldehyde used the bioassay data, it did not incorporate a pharmacokinetic model and the DPX data into OSHA's risk assessments. OSHA asserted that the DPX studies provided qualitative evidence that HCHO could be a carcinogen, but stated that such information could be used with great caution in quantitatively assessing risk. The agency cited the lack of chronic exposure data, a tumor incidence in rats inconsistent with DNA binding results, a paucity of data identifying DNA cross-links, and several other issues as reasons for not incorporating the new studies into its analysis (OSHA, 1987a). As with the CPSC, OSHA had trouble, not with the concept of using the delivered-dose data, but with the methodologies used in the initial CIIT DPX studies.

The most recent OSHA rule on formaldehyde, issued in May 1992 in response to a court remand of the 1987 standard, lowered the standard for occupational exposure to 0.75 ppm (OSHA, 1992). Using the same risk assessment developed in 1987, the agency estimated that from 0.2 to 72 cancers (depending on whether the maximum likelihood estimate or the upper confidence limit was used) would be avoided over the next 45 years with the standard lowered to 0.75 ppm. Again, while there was a substantial cost to industry, the standard was judged to be both economically and tech-nologically feasible.

As in 1987, OSHA again did not include a pharmacokinetic model or the DPX data in its analysis. The agency stated that its reasons behind this omission remain the same as they were in 1987. Moreover, since advocates of the new studies did not claim that they would justify a different standard, the agency did not believe that it was necessary or appropriate to reopen the rulemaking record to include them (OSHA, 1992).

In essence, OSHA has decided that, at least for now, the 1987 risk assessment without the pharmacokinetic model is sufficient. Redoing the risk estimates with the new data would require reopening the rulemaking process. Such an action is financially draining on both the agency and the other parties involved. The agency worked with industry and the relevant unions to resolve the problems with the 1987 standard and to develop a new standard that was economically feasible and satisfactory to all parties. If the agency were pressed into new rulemaking procedures in the future, industry would undoubtedly push for including the pharmacokinetic model and DPX data. While industry has so far been able to accommodate the increasingly restrictive OSHA regulations for HCHO, it would use the new data to defend its position that no further reductions in emissions standards are necessary (personal communication with Formaldehyde Institute, 1992). As it stands, however, industry is anxious to achieve regulatory certainty and seems willing to live with the current risk assessment (personal commu-nication with OSHA, 1992).

Environmental Protection Agency

From the perspective of the new scientific studies and how to interpret them, the regulatory story on formaldehyde is perhaps the most interesting at the Environmental Protection Agency. Under the Toxic Substances Control Act of 1976 (TSCA), the EPA is authorized to regulate chemicals posing an unreasonable risk to health. The agency was slow to act on the initial bioassay data for HCHO, however.

Under Administrator Anne Gorsuch, the agency elected not to designate formaldehyde as a priority chemical under Section 4(f) of TSCA and did not initiate any regulatory activity. In 1983 the agency sponsored a workshop of experts on the chemical in an effort to evaluate the available data and reach some consensus on HCHO. The group failed to resolve key technical disagreements, however, and could not agree on the carcinogenic properties and potential of formaldehyde (Consensus Workshop, 1984). It was not until the following year, after William Ruckelshaus had taken over as administrator and the agency was facing a lawsuit from the Natural Resources Defense Council and the United Auto Workers, that the EPA reconsidered its decision and assigned high priority to HCHO.

The designation of formaldehyde as a priority chemical spurred the agency to take action on several fronts. One of these was the initiation of a continuing regulatory investigation into exposure to HCHO in homes. Over the next several years, risk managers at EPA explored various control options for the chemical and the attending costs for these options.

In conjunction with the risk management strategy, the agency simultaneously developed a quantitative risk assessment of the health effects of formaldehyde, using both the CIIT bioassay data and the available epidemiologic studies. Written by the Office of Toxic Substances (OTS), the 1985 analysis concluded that the chemical should be seen as a carcinogen in rats and, based on epidemiologic data perceived as "limited," may be a human carcinogen (EPA, 1985).

With the high levels of scientific uncertainty surrounding the chemical and its colorful history, the EPA sent the risk assessment document to its extramural review panel, the Science Advisory Board (SAB). One of the key issues of concern to OTS was how to interpret and incorporate the model using the pharmacokinetic data on formaldehyde into the risk assessment process. The agency had chosen not to include this model, but was seeking advice on the issue.

The Environmental Health Committee of the SAB gave the document a generally favorable review. Its response to the issue of incorporating a pharmacokinetic model and data, however, is of particular interest. Several members of the review panel were familiar with the CIIT studies and strongly believed that a model using the data should be incorporated into the analysis (Transcript of Environmental Health Committee meeting, 1985). The SAB recommended that a special review panel, independent of the agency, be formed to evaluate the pharmacokinetic data (Environmental Health Committee, 1985).

In collaboration with a scientific consulting firm, Life Systems, Inc., the EPA convened a panel of seven scientists from a variety of disciplines to review the CIIT pharmacokinetic studies. The group was to review the new data to determine whether they were appropriate both for use in assessing the health risks of formaldehyde and for incorporation into the risk assessment document (Life Systems, Inc., 1986). In January 1986 the panel released its report, which was very critical of the CIIT research. Of particular concern were the assumptions and procedures used in the pharmacokinetic studies, particularly the analytic methods used to measure the tissue dose of the chemical. While recognizing that the work represented an important first step toward the introduction of intracellular dosimetry into the risk assessment process, the report concluded that the research did not provide a basis for quantifying risk (Life Systems, Inc., 1986). CIIT scientists contested the panel's conclusion, to no avail (Casanova et al., 1986).

In 1987 the OTS issued a final risk assessment document (EPA, 1987). Again, a pharmacokinetic model and the DPX data were not included. Although several members of the Science Advisory Board, as well as CIIT scientists, did not agree with the conclusions of the special panel, the EPA heeded the panel's recommendation that the data should not be used for risk quantification. The agency maintained its original position that the use of the data for risk assessment was "premature" (DiCarlo, undated). The agency accepted the 1987 analysis as a working document and shifted formaldehyde over to OTS risk managers to develop policy options for reducing potential residential cancer risks (Putnam, 1991).

Before sending their options analysis to senior EPA decision makers in 1989, however, OTS staff explored what further developments had emerged in formaldehyde science. They wanted to ensure that the office was familiar with the latest data on the chemical. Believing that the recent pharmacokinetic studies were important, the OTS decided that the risk assessment should be updated to include a pharmacokinetic model and the new CIIT data.

In the updated risk assessment document released in 1991, the OTS included both the rat and monkey DPX data. The authors of the analysis determined that the more recent CIIT research had clarified most of the concerns expressed by the previous special panel. They concluded that, although some uncertainties remain about the pharmacokinetic data, their use in risk estimates is appropriate. This decision to incorporate the DPX data was reinforced by remaining uncertainties in the epidemiologic data. While the Blair and other large studies provided more human data than had previously been available, the OTS determined that there were still not enough epidemiologic exposure data to quantify a dose-response relationship for the chemical and, hence, it could not calculate low-dose human cancer risk (EPA, 1991). To make the risk assessment as biologically plausible as possible, the OTS strongly favored the use of the DPX data (personal communication with OTS, 1992). The OTS also recognized the importance of the new cell proliferation data, although the authors determined that they could not yet incorporate these data into their current risk assessment models.

Again the OTS sent its risk assessment document to be reviewed by the Environmental Health Committee of the agency's Science Advisory Board. In an initial consultation (a less formal review), many of the members of the panel reported favorably on the document and its use of a pharmacokinetic model in the analysis. The document and its review were mired in controversy, however. The Science Advisory Board was criticized by the United Auto Workers and other environmental groups as being unbalanced in scientific opinion and institutional representation, and biased in its meeting procedures. Of particular concern to these groups was what was perceived to be a heavy weighting of panel members who favored the use of a pharmacokinetic model and the DPX data, and a lack of members with an epidemiologic background.

The Environmental Health Committee met for formal review of the formaldehyde risk assessment update in July 1991. This time the panel included three new members, each of whom was known to have a conservative view on the use of a pharmacokinetic model in the risk assessment process. The incorporation of such a model and the DPX data was clearly one of the most contentious issues at the meeting, and the panel had a difficult time trying to reach consensus (personal communication with SAB, September 1992).

The SAB report presented a mixed picture. The panel recognized the advances brought by the pharmacokinetic data, but stated that the use of delivered dose measures in quantitative risk assessment remained equivocal, except as a measure of exposure. It also pointed out that because of weaknesses in the epidemiologic data, such as possible confounding with simultaneous exposure to particulates, only point estimates with extensive uncertainties were likely to be available. The report recommended that the risk estimates based on the DPX data be compared with those derived from the most appropriate epidemiologic studies, such as the Blair study. The panel advocated that the risk assessment document should present estimates and models based on rats, monkeys, and humans, and then compare the advantages and disadvantages of each data source and model (Science Advisory Board, 1992).

On the EPA regulatory front, risk managers at OTS realized that the incorporation of the pharmacokinetic model and DPX data into the risk assessment would lower the estimates of human cancer risk for exposure to formaldehyde. This would in turn lower the perceived benefits of strong regulatory action, at least for those exposed to the low levels typically found in homes. The OTS risk managers began to focus on identifying potential risk reduction actions for HCHO oriented to avoiding noncancer human effects in residential settings.

Most recently, while the OTS is working on the latest version of the risk assessment document, risk managers have also been pursuing a regulatory strategy to address the issue of indoor air quality posed by HCHO emissions from urea-formaldehyde pressed wood building materials, cabinets, and furniture. This includes supporting the revision of the emissions standard set by the American National Standards Institute for urea-formaldehyde particleboard flooring products from 0.3 to 0.2 ppm (Greenwood, 1992).

THE STATES ENTER THE RING

The risk assessors at OTS were not the only regulatory proponents of the biologic data for formaldehyde. While the EPA and its Science Advisory Board were struggling over the appropriateness of the data for the agency's risk assessment, the state of California released its own assessment of the health effects of the chemical. This document similarly explored the use of the new biologic data to estimate HCHO's cancer risk to humans.

Like the most recent EPA document, the California analysis used the DNA cross-links (although California used only the rat DPX data) as an indicator of tissue dose. The risk assessment also explored the use of cell proliferation data in developing a range of risk estimates. The authors concluded that the use of these data yielded a more accurate risk assessment (California Air Resources Board, 1992).

The California analysis incorporated a variety of different assumptions than the analysis done by the EPA. One of the key assumptions involved the use of tumors in the analysis: the EPA focused specifically on nasal cancer from the rat data; California tried to account for other cancers in the respiratory tract as well as lung cancer using the rat data and animal-to-human scaling factors (personal communication with California Air Resources Board, 1992). The resultant unit risk for a lifetime of exposure estimated by the analysis—7×10^{-3} ppm^{-1}—was several times higher than those proposed by the federal agency. Moreover, the California analysis used only the rat data, not the monkey data, for the interspecies extrapolation to human risk.

California has been criticized for its interpretation of the new biologic data in its risk assessment. Of particular note is the belief by several CIIT scientists that California misused their data in the risk models. For example, critics have argued that the California model has a strong conservative bias and overpredicts the human risk from formaldehyde. This bias results from such factors as the model's overprediction of DPX levels observed in rats exposed to low formaldehyde concentrations, the interspecies extrapolation procedures used in the model, and the failure to adequately account for formaldehyde's effects on cell replication in target tissues (Starr, 1991). Furthermore, the critics chided the California model for disregarding the monkey DPX data. While EPA's risk assessment was parallel with the models developed by Starr and others at CIIT, California chose to incorporate different assumptions and modeling techniques.

The California Air Resources Board has declared HCHO a toxic air contaminant. The risk assessment has gone to the state's Office of Administrative Law for review. What regulatory action, if any, the state will ultimately pursue remains to be seen.

THE DEBATE

From almost the moment the first pharmacokinetic data were released, they generated controversy in the scientific and regulatory arenas. The key issue was how the results of the studies were to be interpreted for use in quantitative risk assessment. Sides were quickly chosen over whether the data should be used in these analyses to evaluate the risk that formaldehyde poses for humans.

Critics of the early pharmacokinetic data were quick to take issue with many of the methodological and design components of the research. Opponents challenged the relevance of the short-term studies for the risk estimates associated with the chronic exposure experienced by humans. They were also concerned not only about the measurement of the DNA-protein cross-links but also about the relationship of the cross-links to carcinogenesis. The critics argued that this relationship had not been defined for formaldehyde, nor, indeed, for other chemical carcinogens.

Despite the further studies that have been released furnishing additional pharmacokinetic and mechanistic data for the scientific debate, critics remain. Opponents[2] contend that while the science of pharmacokinetics is a reasonable and promising field in principle, the data on formaldehyde are not yet of sufficient quality to permit the incorporation of a pharmacokinetic model into formal risk analyses. They argue that while the scientific methodologies are improving, the DNA adduct and other measurements are not yet strong enough to be included in human risk estimates. Also, given the uncertainties about the data that still remain, the numbers obtained in new analyses using these data may not be any closer to the "truth" than the previous work. Critics contend that if we base risk estimates on the DPX data, and they turn out to be incorrect, we might be seriously underestimating human risk and endangering the health of the population.

Champions of using a pharmacokinetic model and the new data, on the other hand, perceive the additional information as bringing us closer to the truth. They argue that the DPX data significantly improve our understanding of what is happening to the tissues in the body. Because of the naturally occurring defense mechanisms, such as mucociliary systems, there is little covalent binding at the lower exposures. However, as the concentrations increase, the defense mechanisms saturate and the amount of formaldehyde reaching the DNA increases disproportionately. These data yield a more accurate measure of dose, and in turn provide for a more accurate risk assessment. Furthermore, the glutathione and isotope research improved many of the earlier methodological issues and unknowns with the data, the monkey data support the effects seen in the rat and have the potential to serve as a more appropriate species surrogate for humans; and, finally, chronic DPX studies are currently being conducted with results soon to be released.

Proponents of the cell proliferation data similarly profess the importance of the data. They argue that carcinogenesis is a multistage process. Sustained cell proliferation, which fixes unstable adducts into mutations, is necessary to get cancerous growths. Just as the DPX data support the view of a nonlinear relationship between administered and delivered dose for formaldehyde, the cell proliferation data posit a nonlinearity in the dose response curve. Proponents argue that it is important to be as realistic as possible and incorporate this nonlinearity into the risk assessment process, not just rely on the customary defaults of the current risk assessment models—such as the no-threshold, low-dose linear response curve assumed with mutagenesis. While the cell proliferation data are currently difficult to incorporate into the risk assessment process, the champions of the data assert that it is important to develop new models and paradigms that would facilitate their use.

In the regulatory arena, risk assessors may be reluctant to use the new data in their risk analyses. Regulatory agencies have traditionally taken a conservative stance, choosing to err on the side of being overly protective of human health. They may be hesitant to use the new data, knowing that the resultant risk estimates will be smaller than if the data are not incorporated. This lowering of the risk estimate is particularly pertinent to the low-dose exposure levels typically faced by chronic human exposure and traditionally the focus of regulatory policy. Furthermore, risk assessors receive little guidance on what data to use in their analyses. Legislative mandates governing the agencies do not provide many criteria as to when to depart from the traditional models and data sets and incorporate new information into the risk assessment estimates (Rosenthal et al., 1992).

OBSERVATIONS

The formaldehyde case offers a number of interesting observations on the use of new science in the regulatory arena. The first of these observations illustrates an apparent mismatch between the generation of new science and regulatory decisions. As the HCHO case demonstrates, regulatory decisions may be only weakly influenced by new data. The CPSC acted promptly on the initial CIIT bioassay data, but it took a federal court decision for OSHA to finally use the bioassay data in its risk assessment and rulemaking process. Although it is now 14 years since the publication of CIIT's bioassay results, the EPA still lacks a settled regulatory policy on HCHO. None of the three agencies has yet officially incorporated the new findings on HCHO into their risk estimates or regulatory decisions.

The second point of interest centers on the observation that much of the debate about the new information concerns not the validity of the new science, but how to interpret it for risk assessment and regulatory use. For example, some scientists interpret the lower rate of DNA-protein cross-link formation and larger target areas seen in the monkey DPX data (as opposed to the rat) as predicting a lower cancer risk for primates, and, ultimately, a lower one for humans than was estimated using the rat data. Other scientists, however, have interpreted these same data with their larger target area as supporting the observation of upper airway and lung cancer in people seen in the Blair study, in contrast to the nasal cancer seen in the rodent bioassays. This interpretation of the monkey data would lead to a higher cancer risk for humans than that estimated from the rat data.

This matter is further complicated by the problem that several pieces of the new data—from different scientific disciplines—appear to be in opposition to each other. A prevalent interpretation and use of the DPX data lowers the risk estimates for human cancer; a prevalent interpretation of the epidemiologic data raises the risk estimates for human cancer. In essence, use of different pieces of the new information results in contrasting risk estimates. Those on one side of the scientific—and regulatory—debate can cite some of the new scientific information in defense of their views, while those on the other side cite other pieces of the new data.

Unfortunately, the debate over the interpretation and regulatory use of the new science on formaldehyde is also tainted by how the numbers play out in the risk assessment process. It can be argued that because the incorporation of a pharmacokinetic model and the DPX data may lower the risk estimates, red flags were immediately raised in several camps. Those charged with protecting human health in the regulatory arena worried about the new data because they may lower the estimated risk of exposure to formaldehyde. While supporting the use of new science in principle, their job ultimately compels them to oppose lowering the risk estimates if there remains any uncertainty in the interpretation on which they can base their argument. Similarly, industry representatives welcomed the new data as supporting their position that exposure to formaldehyde at current levels is "safe," perhaps even overly protective. Had the numbers played out in the opposite direction, it is possible that the players would have taken different sides in the regulatory debate.

Although the resulting risk estimates should not ultimately affect scientists' perception of the data, it is difficult to imagine that scientists can always divorce themselves from the results of the analyses. In a perfect world, scientists are solely interested in the generation and interpretation of "good" science, regardless of other issues such as their sponsorship or how the numbers play out—which may be the case with many scientists today. Unfortunately, the possibility exists that, while scientists may not be intimately involved in the politics of the regulatory arena, they may have an inherent level of conservatism that dictates their perspective on new data. It is unfair to speculate on potential differences in the scientific debate were the risk assessment numbers to be more conservative using the new data, but it is impossible not to wonder about it.

A third observation involves the source of the new data. Much of the DPX and cell proliferation data on formaldehyde emanated from the Chemical Industry Institute of Toxicology, a research institution supported largely by industry dollars. Not only did the data come from a single institute, but the funds to develop them flowed from industries with a distinct interest in lowering the risk estimates for the chemical. On the one hand, one could argue for more studies from diverse institutions to either refute or corroborate the CIIT data for a more balanced generation of the science. But, on the other hand, the size and cost of these types of studies make such a wish somewhat unrealistic. Because industry has such a large stake in formaldehyde, companies are willing to support major research operations on the chemical. It is unlikely that government agencies would have either the resources or the personnel necessary to undertake such projects.

CIIT scientists are also ardent advocates of using the new science in risk assessment. In many of the papers describing their studies, CIIT researchers fervently champion the incorporation of the study results into the risk assessment and regulatory process. They are proud of the work that they have done on HCHO and would like it to be used to further the regulatory process as well as the scientific environment. Several of the scientists have developed risk assessments themselves using the latest available data, and CIIT scientists frequently attend public hearings (such as SAB meetings) on the data.

A fourth key observation is the level of "politics" involved in incorporating new science into the regulatory arena. This issue is well illustrated by the SAB panel review of the 1991 EPA risk assessment on HCHO.

Initially, in its preliminary review of the document, the panel supported the agency's use of the DPX data. The SAB, however, received much criticism from the UAW and others who opposed the DPX data and who had an important stake in not lowering the human risk estimate. Several additional members were subsequently added to the SAB panel, and, at the new review of the risk assessment, the group was split on whether to incorporate the DPX data.

Also, not only is the clout of powerful interested players of note here, but why the UAW would get so involved with an agency that has only secondary jurisdiction in areas concerning occupational risk. One might speculate that if the debate over the new data were to disappear and the EPA were to use them, then maybe OSHA would similarly feel compelled to incorporate them in its next risk assessment as well. If this were the case, perhaps it would be in the best interest of the labor unions to protect their flanks and to discourage the use of the new data at a variety of points in the regulatory process, not just at OSHA.

CONCLUSION

The path that new scientific information travels from the laboratory to the regulatory arena is neither straightforward nor predictable. The time-table involved is often a long one. Numerous factors, such as scientific uncertainty, technical feasibility, and economic and political viability, con-found and entangle the incorporation of new science into the decision-making process. Furthermore, there are a myriad of players who interact with and complicate the process.

In the simplest or "naive" depiction of this process, new science gets incorporated into agency risk assessments of the relevant risk. These risk assessments then become the basis for regulatory decisions (see Figure 4). In an ideal world, it would be nice if the regulatory process were as simple and efficient. Regulatory reality, however, tells a different story. There are numerous perturbations that can disrupt and convolute this simple model.

The first complication to the model arises from the fact that science can bypass the risk assessment process and directly influence regulatory deci-sions (Figure 5). The scientific data, such as the HCHO epidemiologic data, may be difficult to quantify and incorporate into a formal risk assess-ment. While risk assessors may disagree over the quality of the science and its acceptability for their analyses, the mere existence of the new science may perturb risk managers to worry about the "truth" of the issue at hand. Moreover, the development of new science may serve to slow down the regulatory decision-making process, while risk assessors and risk managers debate its regulatory appropriateness.

The regulatory model can be further complicated by the role that addi-

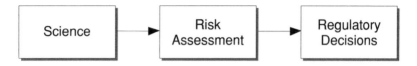

FIGURE 4   Naive model of regulatory process.

tional groups play in the process. One key player to interact in the model is the courts (Figure 6). The courts can interact with the regulatory decision-making process by influencing either the risk assessment stage or regulatory decisions. Many regulatory actions—or inactions—frequently land in the courts, where judges ultimately decide their legality and appropriateness. The formaldehyde case demonstrates the power of the courts on several fronts. For example, the federal district court forced future regulatory action by ordering OSHA to consider formally initiating a new rulemaking process for formaldehyde, in response to a suit brought against the agency by several groups seeking an emergency temporary standard for the chemical. On the other hand, the court of appeals overturned CPSC's 1982 ban on urea-formaldehyde foam insulation, ruling the CPSC had failed to support its action with substantial evidence. As illustrated in the HCHO case, the courts can be either a compelling or an inhibiting force in the regulatory process.

Another key influence in the process is the prevailing presidential philosophy. This philosophy can impact the process again at either the risk assessment or decision-making stage (Figure 7). It can also impact the courts through presidential appointment of judges. The influence of presidential philosophy is clearly depicted in the formaldehyde case through the "regulatory reduction" ideology of the Reagan administration. At both OSHA and EPA, agency leaders following this ideology were slow to act on HCHO in the early 1980s. It took powerful influences—the courts in the case of OSHA, agency turmoil resulting in a new administrator in the case of EPA— to spur regulatory action at the agencies. The Republican antiregulatory

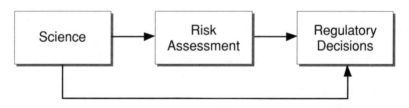

FIGURE 5   Model of regulatory process: Complication #1.

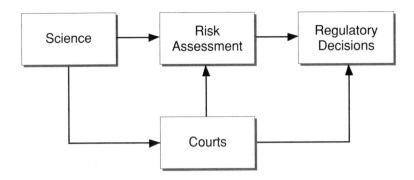

FIGURE 6   Model of regulatory process: Complication #2.

philosophy further permeated the process, however, by requiring that the Office of Management and Budget (and later the Council on Competitiveness) approve potential agency regulations.

A fourth complication to the naive regulatory model is presented by the powerful influence of various interest groups. These groups, such as industry associations or public interest advocates, can impact the model at nu-

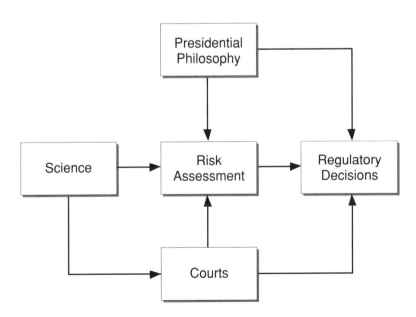

FIGURE 7   Model of regulatory process: Complication #3.

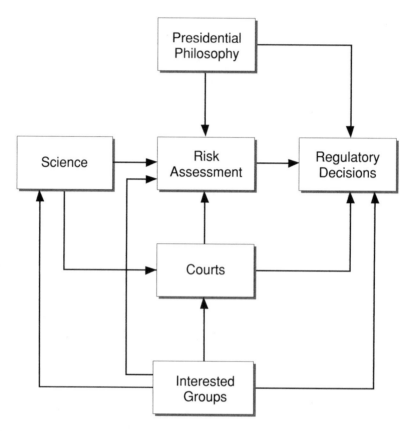

FIGURE 8   Model of regulatory process: Complication #4.

merous points (Figure 8). The formaldehyde case illustrates the influence
of these groups at many junctures. CIIT, an industry-sponsored research
institute, generated both the initial animal bioassay and subsequent DPX
data on formaldehyde. Furthermore, CIIT scientists not only adamantly
advocated the incorporation of these data into agency risk assessments but
also performed and published their own risk assessments using the data.
Other groups, such as the United Auto Workers and several environmental
groups, took OSHA to court over its initial inaction on HCHO. These groups,
as well as the Formaldehyde Institute, again sued the agency over its 1987
regulatory decision (the 1.0 ppm standard) on the chemical. At EPA, both
industry and labor groups lobbied the agency on the relevant data to incorporate into the risk assessment process and gave testimony to the agency's Science Advisory Board review of formaldehyde. At every step in the formaldehyde story, powerful interest groups have exerted their influence.

Perhaps the ultimate issue illustrated by these models and their increasing complications is, Who decides when new science should be incorporated into the regulatory process? Who converts science on the "frontier" into "settled" science, acceptable for risk assessment and regulatory decisions? Is it the agencies—the risk assessors, the risk managers? Is it an external panel of scientists who reach some consensus on the data's acceptability? Should Congress legislate what scientific methodologies or data sets must be used? Should it be left up to the courts to decide? There will always be debate over what science is acceptable or what methodologies are valid. Furthermore, as the new OSHA standards for formaldehyde illustrate, regulatory actions do not come cheaply; there are significant economic as well as health issues involved in using or not using new scientific information. The issue ultimately becomes twofold. First, who is qualified to make these judgments? Second, who will be held accountable for the consequences of these judgments?

NOTES

1. When cells are exposed to formaldehyde, there may occur a cross-linking between DNA and proteins in the cells. These cross-links can then be counted as a measure of the HCHO dose that has actually reached the cells—the delivered dose (as opposed to the administered dose).

2. The views expressed here on the most recent formaldehyde biologic data by both opponents and proponents of the data were gathered from informal personal communications with a number of scientists knowledgeable on the issue, many of whom wished their comments to remain anonymous. This includes comments on the new data from some of the panel of scientists originally interviewed for their opinion on the initial pharmacokinetic data (Casanova-Schmitz et al., 1984) in Hawkins and Graham (1988).

REFERENCES

Albert, R. E., A. R. Sellakmur, S. Laskin, M. Kuschner, N. Nelson, and C. A. Snyder. 1982. Gaseous formaldehyde and hydrogen chloride induction of nasal cancer in the rat. Journal of the National Cancer Institute 68:597–603.

Blair, A., P. Stewart, M. O'Berg, W. Gaffey, J. Walrath, J. Ward, R. Bales, S. Kaplan, and D. Cubit. 1986. Mortality among industrial workers exposed to formaldehyde. Journal of the National Cancer Institute 76:1071–1084.

Blair, A., P. A. Stewart, R. N. Hoover, J. F. Fraumeni, Jr., J. Walrath, M. O'Berg, and W. Gaffey. 1987. Cancers of the nasopharynx and oropharynx and formaldehyde exposure (letter to the editor). Journal of the National Cancer Institute 78:191.

Blair, A., P. A. Stewart, and R. N. Hoover. 1990. Mortality from lung cancer among workers employed in formaldehyde industries. American Journal of Industrial Medicine 17:683–699.

California Air Resources Board, Office of Environmental Health Hazard Assessment. 1992. Cancer Risk Assessment for Airborne Formaldehyde. Proposed Identification of Formaldehyde as a Toxic Air Contaminant: Part B, Health Assessment. Sacramento, Calif.:California Air Resources Board.

Casanova-Schmitz, M., T. B. Starr, and H. d'A. Heck. 1984. Differentiation between meta-bolic incorporation and covalent binding in the labeling of macromolecules in the rat nasal mucosa and bone marrow by inhaled [^{14}C]- and [^{3}H] formaldehyde. Toxicology and Applied Pharmacology 76:26–44.

Casanova, M., T. B. Starr, and H. d'A. Heck. 1986. Comments on the final report of the panel reviewing the CIIT pharmacokinetic data on formaldehyde. Research Triangle Park, N.C.

Casanova, M., and H. d'A. Heck. 1987. Further studies of the metabolic incorporation and covalent binding of inhaled [^{3}H]- and [^{14}C] formaldehyde in Fischer-344 rats: Effects of glutathione depletion. Toxicology and Applied Pharmacology 89:105–121.

Casanova, M., D. F. Deyo, and H. d'A. Heck. 1989. Covalent binding of inhaled formalde-hyde to DNA in the nasal mucosa of Fischer 344 rats: Analysis of formaldehyde and DNA by high-performance liquid chromatography and provisional pharmacokinetic inter-pretation. Fundamental and Applied Toxicology 12:297–417.

Casanova, M., K. T. Morgan, W. H. Steinhagen, J. I. Everitt, J. A. Popp, and H. d'A. Heck. 1991. Covalent binding of inhaled formaldehyde to DNA in the respiratory tract of Rhesus monkeys: Pharmacokinetics, rat-to-monkey interspecies scaling, and extrapola-tion to man. Fundamental and Applied Toxicology 17:409–428.

Cascieri, T. C., and J. J. Clary. 1992. Formaldehyde-oral toxicity assessment. Comments in Toxicology 4:295–304.

Cohn, M. S., F. J. DiCarlo, A. Turturro, and A. G. Ulsamer. 1985. Letter to the editor. Toxicology and Applied Pharmacology 77:363–364.

Collins, J. J., J. C. Caporossi, and H. M. D. Utidjian. 1987. Letter to the editor. Journal of the National Cancer Institute 78:192–193.

Collins, J. J., J. C. Caporossi, H. M. D. Utidjian. 1988. Formaldehyde exposure and nasopha-ryngeal cancer: Re-examination of the National Cancer Institute Study and an update of one plant (letter to the editor). Journal of the National Cancer Institute 80:376–377.

Consensus Workshop on Formaldehyde. 1984. Report on the Consensus Conference on Formaldehyde. Environmental Health Perspectives 58:323–381.

Consumer Product Safety Commission. 1982. Ban of urea-formaldehyde foam insulation. Federal Register 47(April 2):14366–14419.

DiCarlo, F. undated. Memorandum on the Expert Review of CIIT Pharmacokinetic Data on Formaldehyde. Washington, D.C.: U.S. Environmental Protection Agency.

Environmental Health Committee. 1985. Review Report of the Draft Document, Preliminary Assessment of Health Risks to Garment Workers and Certain Home Residents from Ex-posures to Formaldehyde. Washington, D.C.: U.S. EPA Science Advisory Board.

Federal Panel on Formaldehyde. 1982. Report of the Federal Panel on Formaldehyde. Envi-ronmental Health Perspectives 43:139–168.

Gibson, J. E., ed. 1983. Formaldehyde Toxicity. Washington, D.C.:Hemisphere Publishing Corp.

Goldmacher, V. S., and W. G. Thilly. 1983. Formaldehyde is mutagenic for cultured human cells. Mutation Research 116:417–422.

Graham, J. D., L. C. Green, and M. J. Roberts. 1988. In Search of Safety. Cambridge, Mass.: Harvard University Press.

Greenwood, M. A., Office of Prevention, Pesticides, and Toxic Substances, U.S. EPA. 1992. Letter to William McCredie at the National Particleboard Association, July.

Gulf South Insulation v. Consumer Product Safety Commission. 701 Fed.2d. 5th Circuit. 1137 (1983).

Hawkins, N. C., and J. D. Graham. 1988. Expert scientific judgment and cancer risk assess-ment: A pilot study of pharmacokinetic data. Risk Analysis 4:615–625.

Heck, H. d'A., and M. Casanova. 1987. Isotope effects and their implications for the covalent binding of inhaled [^{3}H]- and [^{14}C] formaldehyde in the rat nasal mucosa. Toxicology and Applied Pharmacology 89:122–134.

Jasanoff, S. 1990. The Fifth Branch. Cambridge, Mass.: Harvard University Press.

Kerns, W. D., K. L. Pavkov, D. J. Donofrio, E. J. Gralla, and J. A. Swenberg. 1983. Carcinogenicity of formaldehyde in rats and mice after long-term inhalation exposure. Cancer Research 43:4382–4392.

Life Systems, Inc. January 1986. Expert Review of Pharmacokinetic Data: Formaldehyde. Washington, D.C.

Monticello, T. N. 1990. Formaldehyde-induced pathology and cell proliferation. Ph.D. Dissertation submitted to the Department of Pathology, Duke University, North Carolina.

Monticello, T. N., and K. T. Morgan. 1990. Correlation of cell proliferation and inflammation with nasal tumors in F-344 rats following chronic formaldehyde exposure. American Association of Cancer Research 31:139.

Monticello, T. M., K. T. Morgan, J. I. Everitt, and J. A. Popp. 1989. Effects of formaldehyde gas on the respiratory tract of rhesus monkeys. American Journal of Pathology 134:515–527.

Monticello, T. N., F. J. Miller, and K. T. Morgan. 1991. Regional increases in rat nasal epithelial cell proliferation following acute and subchronic inhalation of formaldehyde. Toxicology and Applied Pharmacology 111:409–421.

National Research Council. 1980. Formaldehyde—An Assessment of Its Health Effects. Washington, D.C.: National Academy Press.

Occupational Safety and Health Administration. 1985. Occupational exposure to formaldehyde. Federal Register, vol. 50, no. 74, pp. 15179–15184, April 17, 1985. (To be codified in 29 CFR Part 1910.)

Occupational Safety and Health Administration. 1987a. Conclusions regarding the pharmacokinetic model. Federal Register, vol. 52, pp. 46226–46227.

Occupational Safety and Health Administration. 1987b. Occupational exposure to formaldehyde. Federal Register, vol. 52, no. 233, pp. 46168–46171, December 4, 1987. (To be codified in 29 CFR Parts 1910 and 1926.)

Occupational Safety and Health Administration. 1987c. Summary of the regulatory impact and regulatory flexibility analysis. Federal Register, vol. 52, pp. 46237–46242.

Occupational Safety and Health Administration. 1992. Summary and explanation of the final amendments. Federal Register, vol 57, no. 102, pp. 22292–22328, May 27, 1992. (To be codified in 29 CFR Part 1910.)

Putnam, S. W. 1991. Formaldehyde. Pp. 127–158 in Harnessing Science for Environmental Regulation, J. D. Graham, ed. New York: Praeger Publishers.

Rosenthal, A., G. M. Gray, and J. D. Graham. 1992. Legislating acceptable cancer risk from exposure to toxic chemicals. Ecology Law Quarterly 19:269–362.

Science Advisory Board, U.S. Environmental Protection Agency. 1992. Review of the Office of Toxic Substances' Draft Formaldehyde Risk Assessment Update by the Environmental Health Committee. September. Washington, D.C.: U.S. Environmental Protection Agency.

Starr, T. B. 1990. Quantitative cancer risk estimation for formaldehyde. Risk Analysis 10:85–91.

Starr, T. B. November 1991. Comments on the California Air Resources Board proposed identification of formaldehyde as a toxic air contaminant, September 1991, Draft SRP Version.

Starr, T. B., and R. D. Buck. 1984. The importance of delivered dose in estimating low-dose cancer risk from inhalation exposure to formaldehyde. Fundamental and Applied Toxicology 4:740–753.

Stayner, L. T., L. Elliott, L. Blade, R. Keenlyside, and W. Halperin. 1988. A retrospective cohort mortality study of workers exposed to formaldehyde in the garment industry. American Journal of Industrial Medicine 13:667–681.

Swenberg, J. A., E. A. Gross, and H. W. Randall. 1986. Localization and quantitation of cell proliferation following exposure to nasal irritants. Pp. 291–298 in Toxicology of the Nasal Passages, C. S. Barrow, ed. New York: Hemisphere Publishing Corporation.

Tobe, M., T. Kaneko, Y. Uchida, E. Kamata, Y. Ogawa, Y. Ikeda, and M. Saito. 1985. Studies of the Inhalation Toxicity of Formaldehyde. Report No. TR-85-0236. Tokyo: National Sanitary and Medical Laboratory Service.

Transcript of the Environmental Health Committee Review of the Draft Risk Assessment Document on Formaldehyde. June 26, 1985. U.S. EPA Science Advisory Board.

U.S. Environmental Protection Agency. 1985. Preliminary Assessment of Health Risks to Garment Workers and Certain Home Residents from Exposure to Formaldehyde. Washington, D.C.: U.S. Environmental Protection Agency. Draft.

U.S. Environmental Protection Agency. 1987. Assessment of Health Risks to Garment Workers and Certain Home Residents from Exposure to Formaldehyde. Washington, D.C.: U.S. Environmental Protection Agency.

U.S. Environmental Protection Agency. 1990. Formaldehyde Risk Assessment Update. Washington, D.C.: U.S. Environmental Protection Agency.

U.S. Environmental Protection Agency. 1991. Formaldehyde Risk Assessment Update. Washington, D.C.: U.S. Environmental Protection Agency.

U.S. Regulatory Council. 1979. Statement on regulation of chemical carcinogens. Federal Register, vol. 44, p. 60038.

Vaughan, T. L., C. Strader, S. Davis, and J. R. Daling. 1986. Formaldehyde and cancers of the pharynx, sinus, and nasal cavity: I. Occupational exposures and II. Residential exposures. International Journal of Cancer 38:677–688.

Keeping Pace with Science and Engineering. 1993.
Pp. 221–242. Washington, DC: National Academy Press.

The Dioxin TCDD: A Selective Study of Science and Policy Interaction

John A. Moore, Renate D. Kimbrough, and Michael Gough

The term *dioxin* refers to any one of 75 chemicals that have the same basic chemical structure but vary in the number and location of their chlorine atoms. This case study focuses only on TCDD, which is shorthand for the prototypical dioxin, 2,3,7,8-tetrachlorodibenzo-*p*-dioxin (Figure 1). Related compounds, such as other dioxins or dibenzofurans (a structurally related class of chemicals) did not influence the issues highlighted in this paper and are omitted from most of the discussion. Current scientific thought is that dioxin and dibenzofurans may exert their biological (toxicological) effects through a common mechanism.

This paper briefly describes three activities that involved TCDD to illustrate the interface between scientific information and policy decisions. These are the registration of 2,4,5T for use as a pesticide; the military use of Agent Orange and the Veterans Administration (VA) activities relevant to that use; and setting state water quality standards for TCDD in the Clean Water Act. Figure 2 summarizes the sequence of events.

MEDICAL AND SCIENTIFIC DATA

The dioxin TCDD is not a commercial product. It is a contaminant formed in the manufacture of certain commercial chemicals and a product of the combustion of certain materials. Its toxic potential was first reported by Schultz in 1957. He identified TCDD as the causative agent in the development of a skin disease, chloracne, in chemical workers involved in the production of 2,4,5-trichlorophenol.

221

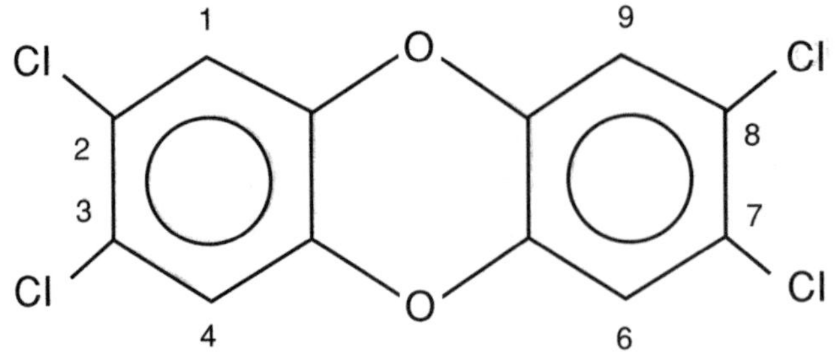

FIGURE 1   Basic chemical structure of 2,3,7,8-tetrachlorodibenzo-*p*-dioxin.

Trichlorophenol was a feedstock for the synthesis of other chemicals such as hexachlorophene and the herbicide 2,4,5-T (2,4,5 trichlorophenoxyacetic acid).   The amount of TCDD formed during the production of 2,4,5 trichlorophenol depended on the chemicals used in synthesis and the temperature of the process; some TCDD levels were in the range of 1 to 30 parts per million.   The amount of TCDD contained as inpurities in hexachlorophene and 2,4,5-T depends on how much was present in the original trichlorophenol. Because hexachlorophene had medicinal uses, a purified form of trichlorophenol was used in its synthesis and TCDD concentrations were in the range of only a few parts per billion.

Animal Studies

Concern about the health effects of TCDD remained narrowly focused on relationships between occupational exposures and chloracne until 1970 when teratology studies with 2,4,5-T indicated that exposure to pregnant mice resulted in the production of cleft palates in offspring (Courtney et al., 1970).   Additional studies (Courtney and Moore, 1971; Sparschu et al., 1971) soon revealed that it was the TCDD impurity in 2,4,5-T that accounted for the vast majority of the teratogenic or fetotoxic responses in mice and rats.   In subsequent years a number of studies confirmed and extended the developmental toxicity observed with TCDD.   In addition, studies with TCDD that spanned three generations reported adverse reproductive effects (Murray et al., 1979).

Several years later, a number of male veterans who served in Vietnam expressed concern that birth defects in their children were a consequence of their military exposure to Agent Orange, a mixture of the herbicides 2,4,-D (2,4 dichlorophenoxyacetic acid) and 2,4,5-T. To investigate the possibility of a causal association between exposure to Agent Orange and birth defects, an extensive study was performed in male mice.   In these studies, high doses of Agent Orange, and TCDD, were employed—sufficient to cause mild toxicity in the mice.   Nonetheless, fertility was not impaired and no

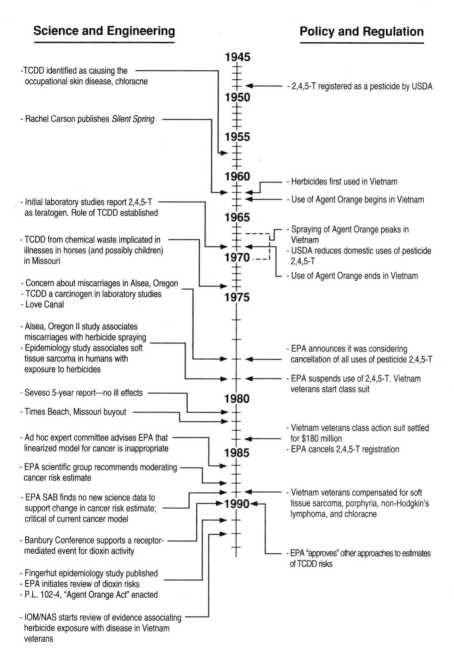

FIGURE 2   Timeline of significant scientific, societal, and regulatory events involving the dioxin TCDD.

association with birth defects was discovered, since the incidence of offspring with birth defects was similar to that seen with untreated mice (Lamb et al., 1981a,b).

Another important health study was reported in 1978. In a long-term study in rats, an increased incidence of tumors was associated with TCDD exposure (Kociba et al., 1978). These results were soon confirmed by other studies in mice and rats; currently there are well over a dozen research publications associating TCDD with the development of cancer in animals; it is generally held that TCDD "promotes," as opposed to "initiates," the expression of a carcinogenic response, probably through some form of hormonal interaction.

Epidemiological Studies

Epidemiological studies of the effects of TCDD in humans are almost exclusively investigations in males. For example, there are no definitive human studies of reproductive and developmental effects in women exposed to TCDD. The one study of exposures to women in and around Seveso, Italy, after TCDD contamination, did not demonstrate an increased risk of birth defects (Mastroiacovo et al., 1988). However, the statistical power of the finding was limited by the size of the study population.

Most of the studies done to date have focused on male Vietnam veterans, the Air Force personnel that applied Agent Orange in Vietnam (Operation Ranch Hand), or workers whose occupation(s) resulted in exposure to TCDD.

In response to veterans' concerns about birth defects, the Centers for Disease Control (CDC) undertook a case control study using data from the Metropolitan Atlanta Congenital Birth Defects Program. This CDC birth defects study was designed to determine if Vietnam veterans were fathering children with a higher incidence of birth defects. The potential risk factor being investigated was service in Vietnam, not TCDD exposure, per se, since there were no records that allowed such a distinction. The study results did not establish an association between service in Vietnam and a higher incidence of birth defects in children (Erickson et al., 1984).

The Ranch Hand studies, which are continuing, are the most thorough in breadth and frequency of medical surveillance. The study focuses on the 1,200 persons that were part of the Ranch Hand operation, ranging from pilots and crew that flew the missions to the ground personnel that serviced the aircraft and handled the storage and loading of Agent Orange. These individuals periodically receive a thorough medical examination, and fertility is followed as are incidence of disease and cause of death. Thus far, these studies show no patterns that indicate a health detriment due to TCDD exposure. Significant associations between serum dioxin levels and variables related to lipids (fats that are not soluble in water) were found. Since TCDD is soluble in lipids, the association may reflect a difference in the pharmacokinetics of TCDD in different subsets of the Ranch Hand population.

Many individuals in the Ranch Hand study were found to have elevated TCDD levels in their blood serum (Air Force Health Study, 1991), providing objective evidence of the degree of past exposure to Agent Orange or possibly other herbicides containing TCDD that were used in comparatively small quantities in Vietnam. Other researchers have reported increased TCDD levels in adipose tissue (fatty tissue) in veterans who were associated with herbicide application in Vietnam (Schecter et al., 1989). In contrast, studies of veterans of ground warfare in Vietnam have not reported increased levels of dioxin (CDC, 1988).

In the absence of an identifiable group of ground warfare veterans exposed to TCDD, several studies have compared the health of Vietnam veterans with that of other veterans. One of those studies reported a service-related increase in non-Hodgkin's lymphoma, which is unrelated to exposure to TCDD (Selected Cancers Cooperative Study Group, 1990). This increase in incidence was higher among men in the sea-based Navy, a group that is not believed to have had exposure to either Agent Orange or TCDD. No significant adverse effects have been reported in other studies (CDC, 1988; Kahn et al., 1988; Kang et al., 1991; Gough, 1991).

Hardell and coworkers (Ericksson et al., 1981; Ericksson et al., 1990; Hardell et al., 1979; Hardell, 1981; Hardell et al., 1981) studied agricultural and forestry workers and reported a positive association between exposure to chlorinated phenols or phenoxy acid herbicides and an increased incidence of soft tissue sarcoma or non-Hodgkin's lymphoma. Others have failed to confirm these findings in agricultural and forestry workers (Woods, 1987). The reports of Hardell and coworkers are most directly contradicted by an examination of TCDD levels in New Zealand herbicide sprayers. Although the New Zealand men have elevated levels of TCDD, there is no evidence of any health effects resulting from their exposure to herbicide. It was reported that dioxin levels in patients who were part of Hardell's studies were similar to those in control patients who were undergoing gall bladder surgery (Rappe et al., 1984) Another publication from the same group of Swedish investigators reported that there was no difference in the levels of TCDD or the pattern of other chlorinated dibenzodioxins and dibenzofurans in adipose samples collected from cancer (soft tissue sarcomas or lymphomas) patients classified as "exposed" to phenoxy herbicides and a group of controls that had no history of phenoxy acid herbicides (Nygren et al., 1986). The authors of the New Zealand study conclude that the associations reported by Hardell et al. are almost certainly in error (Smith et al., 1992).

A combined mortality study of chemical workers from twelve different

U.S. plants reported an increase in total tumors, and tumors of the lung and bladder in workers exposed to TCDD and many other chemicals (Fingerhut et al., 1991). An increase in soft tissue sarcomas was also observed in a subset of these workers. In highly exposed workers who became ill long after exposure (20 years), the mortality for all cancers combined was 46 percent higher than expected. This increase was primarily due to a 42 percent increase in cancers of the respiratory tract. Although the increased death rate from total cancers is dominated by cancers of the respiratory tract, very little is known about smoking in the study population. A survey of smoking habits among "surviving workers at only two plants" contributes no information about the smoking habits of workers who had died at all twelve plants studied and who are the subjects of this study. Furthermore, two workers died from mesothelioma, clearly showing that some workers were exposed to other causes of lung cancer, that is, asbestos. Some of the workers who developed bladder cancer were also exposed to chemicals that are considered to be carcinogenic for the human bladder.

Five of the seven workers who developed soft tissue sarcomas had worked in only one of the 12 plants included in the study by Fingerhut and coworkers. In workers from that plant who had soft tissue sarcoma, the effects of exposure to TCDD were confounded by exposure to the known human bladder carcinogen para-aminobiphenyl (International Agency for Research on Cancer, 1972). None of the workers with soft tissue sarcoma were exposed only to TCDD (Collins et al., 1993); all were also exposed to para-aminobiphenyl. These results need more in-depth study. In support of the contention of Collins et al., a study of 1,600 German chemical workers exposed to TCDD found no soft tissue sarcomas (Zober, 1990; Manz et al., 1991). The German study did report an increase in deaths from all cancers, and respiratory cancers are a major contributor to the overall excess of cancer deaths. So far as soft tissue sarcomas are concerned, the two studies contradict each other. Another factor that may confound the interpretation of data on soft tissue sarcoma are errors that are known to occur in the (mis)classification of this tumor type.

All in all, a critical but fair reading of the epidemiology studies leads to the conclusion that deaths from some cancers were elevated in chemical plant workers with unknown smoking habits who were exposed to a variety of chemicals. Among those chemicals was TCDD.

Immunology

Experimental animal studies have shown other toxic effects of TCDD, particularly suppression of immune system responses. Although attempts have been made to detect similar effects in humans, causal associations between TCDD exposure and altered immune status have not been made.

No consistent effects have been observed in studies of Air Force personnel who applied Agent Orange (Lathrop et al., 1984; Lathrop et al., 1987; Wolfe et al., 1991), individuals in Seveso, Italy, who developed chloracne (Mocarelli et al., 1986; Tognoni and Bonaccorsi, 1982), or in a Missouri study of persons who had documented exposure to TCDD (Evans et al., 1988; Webb et al., 1989). These three studies are of particular value since actual body burdens of TCDD could be quantified or else chloracne was present in the study participants. To date, immunotoxic effects have not been a factor in risk management decisions.

REGULATORY LIMITS

For several decades it has been traditional practice in the United States and the developed countries of the world to set quantitative limits on human exposure to a toxic agent by applying a numerical safety factor to a no-observed-effect-level (NOEL). The NOEL is the highest dose in the most sensitive test species at which no toxic effects are observed. In the NOEL-plus-safety-factor approach, exposures below this level are assumed not to pose a risk because one is below the threshold of toxicity.

In 1976 the U.S. Environmental Protection Agency (EPA) changed this practice in estimating cancer risk in the United States on the assumption that there are no NOELs for carcinogens but that risk decreases with dose, reaching zero risk at zero dose. The choice of this specific risk estimation model called the no-threshold model, was strongly influenced by studies of cancer in experimental animals exposed to radiation, the best source of data available at that time. Adoption of this assumption provides numerical estimates of the number of people who may be at risk of cancer in specific situations. These estimates presumably have greater utility than the older procedures, which could identify the number of people exposed above a "safe" level but could not provide the "estimated incidence" of disease.

Risk Models

The "interim" cancer assessment procedures the EPA adopted in 1976 specified the use of linear, no-threshold models. In 1980 the EPA developed an upper estimate of the human cancer risk posed by TCDD using the rat bioassay data of Kociba et al., (1978). This estimate was modestly revised in 1985. The 1985 estimate associates a dose of 0.006 picogram/kilogram (pg/kg)/day with a maximum plausible increased cancer risk of one additional case in a million people.

The EPA Risk Assessment Guidelines for Cancer, which were issued in 1986, retained the 1977 "interim" policy in that the guidelines affirmed the linearized multistage model as the default methodology for estimating cancer risk. Numerical estimates of cancer potency derived from this model are used by EPA in various activities such as setting water quality standards and cleanup levels for hazardous waste sites, and in estimating cancer risk from effluent or ash produced by municipal waste incinerators. Other federal agencies, such as the Food and Drug Administration and the Centers for Disease Control and Prevention, have also derived estimates of TCDD cancer risk using the same basic methods but, owing to differences in other scientific assumptions such as how to adjust for differences in size of mice, rats, and humans, their estimates have produced values that permit exposure levels that are, respectively, nine and four times higher than the EPA's estimate.

Other countries have not elected to apply a linear multistage model as the standard means for estimating cancer risk. Rather, these countries evaluate risks on a case-by-case basis. The practical result of this approach is that the linearized model is used for chemicals that cause genetic toxicity and the NOEL-safety factor approach is used for all other chemicals. Since TCDD is widely interpreted by scientists not to be genotoxic (Zeiger, 1989), the latter approach has been applied to TCDD data, resulting in acceptable exposure limits that are several orders of magnitude greater than those derived by the U.S. EPA.

TCDD as an Exception

Since the mid-1980s, researchers have produced data that indicate the cancer response (and other toxic effects) to TCDD exposure is mediated through binding to a specific molecule, the Ah (aryl hydrocarbon hydroxylase) receptor, in the cells of mammals. Research is currently focused on understanding the molecular events associated with binding, whether there are significant differences among animal species, and the effects that exposure to other chemicals (dioxins, dibenzofurans, polychlorinated biphenyls [PCBs]) may have on the binding of TCDD. In addition, there are differences in pharmacokinetics between humans and animals (Kimbrough, 1990). A basic understanding of these processes is needed to gain an understanding of how receptor binding may be relevant to human risk assessment.

Some scientists have postulated that since TCDD does not cause genetic damage, it cannot "initiate" the essential and irreversible first step in the carcinogenic process. TCDD "promotes" rather than "initiates" tumors. It is plausible to conclude that TCDD affects the carcinogenic process through mechanisms that have a threshold; this makes the standard linearized, multistage model inappropriate for estimating cancer potency. An ad hoc EPA committee in 1986 produced the opinion that the linear model was inappropriate for TCDD. In 1988, a group of senior EPA scientists concluded that "reliance on the linearized model may be less appropriate for TCDD than for many other chemicals, . . . and the model may overestimate the upper bound on the risk by some unknown amount" (EPA, 1988). The group was of the opinion that since TCDD represents less of a toxic risk than previously thought, a higher exposure level could be tolerated without increased risk. The group recommended that the risk-specific-dose (the dose which may result in a one in a million increase in cancer) be raised from 0.006 to 0.1 pg/kg/day. EPA management wanted the opinion of independent experts before it acted on the recommendation and submitted the report to its Science Advisory Board for comment.

The EPA Science Advisory Board established an ad hoc committee to peer review the EPA document, and it agreed that the current linearized multistage model was inappropriate and urged the use of other available models; however, it also stated that there were no new scientific data to support a modification of the value developed in the 1985 risk assessment. The Executive Committee of the EPA Science Advisory Board, in transmitting the peer review comments to the administrator, took the unusual step of expressing their opinion "that the existing LMS-based risk assessment . . . lacks a firm scientific foundation" (EPA, 1988). In a separate activity, many scientists that were experts on the effects of dioxin expressed the opinion at a 1990 Banbury Conference that using traditional models for assessing cancer risk from dioxin is not appropriate; there was consensus that the known toxic effects of dioxin are receptor mediated.

The Current EPA Reassessment

On April 8, 1991, the EPA administrator announced that the agency would reassess the risks of exposure to TCDD and related compounds in light of significant advances that had occurred in the scientific understanding of the mechanisms by which dioxin becomes toxic, the health effects in animals and people, the pathways to human exposure, and the toxic effects of dioxin in the environment. The scientific reassessment of "dioxin" is notable because it does the following things:

1. It broadens the range of scientific issues to be considered from those dominated by cancer to include a wider consideration of such key issues as exposure, dose-response relationship, pharmacokinetics, mechanism of action, immunology, and reproductive and developmental toxicity.

2. It commits the EPA to the development of a biologically based, dose-response model to estimate human health risks. This responds to opinions expressed by scientists since 1986 that the traditional linear multistage model was inappropriate for estimating human risk from TCDD

exposure and implements the recommendation of the EPA Science Advisory Board.

3. It defines a process that enhances the opportunity for the development of a broad scientific consensus. For example, it enlists the participation of a number of the country's dioxin researchers in the review of each topic. A group of scientists outside the EPA reviewed and received comments on the draft documents and will direct the drafting of a risk characterization statement based on these reviews. The latter step is unprecedented in the history of the EPA. The revised scientific chapters and the risk characterization draft will be submitted to the EPA Science Advisory Board for review and comment.

4. It proposes to assess the toxicity potential of not just TCDD but related compounds such as other dioxins, certain dibenzofurans, and selected polychlorinated biphenyls. These other chemicals are believed by some to exert their toxicity through a similar mechanism. Therefore, quantitative estimates of total risk should reflect the sum of exposures to all of these chemicals. The use of this "toxic equivalency" for dioxins and dibenzofurans was reviewed by the EPA Science Advisory Board several years ago and found acceptable as an interim measure.

Extending the toxic equivalency approach beyond dioxins and dibenzofurans to selected PCBs is more controversial, however, because of concerns that the data are inadequate. For example, in 1990 an EPA workshop on PCBs, principally attended by scientists, advised against adopting such a scheme for PCBs because of large data gaps (Barnes et al., 1991). Research data collected since the 1990 workshop, some of which were developed by EPA, indicate that the equivalency values for some PCB congeners (i.e., other chemicals in the same group) are inaccurate by more than two orders of magnitude or incorrectly predict for certain toxic effects, such as cancer potency.

The dioxin reassessment was originally scheduled to be completed in the latter part of 1992; current estimates are that the revised assessment will be submitted to the EPA Science Advisory Board in the spring of 1993 and, in turn, their review will be transmitted to the administrator in late 1993. It is reasonable to expect that completion of the reassessment, irrespective of its content, will have significant implications for science policy and risk management policy.

THREE CASES OF RISK REGULATION

To illuminate the interplay between new information on toxicity and public policy regarding regulation of risk, we explore 2,4,5-T, Agent Orange, and the Clean Water Act in greater detail.

The Pesticide 2,4,5-T

The chemical 2,4,5-T was first registered for use as a herbicide in 1949 and quickly gained wide domestic use for controlling a variety of broad-leafed weeds and woody plants. The registered uses permitted its application in a wide variety of homeowner and agricultural circumstances that included croplands, forestry, and a diverse array of brush control practices, including roadsides, railroad and electrical power line rights of way, and rangeland.

The data indicating TCDD's potential for causing birth defects and other forms of developmental abnormalities in laboratory studies led to the 1970 decision by the U.S. Department of Agriculture, the agency that then regulated pesticides in the United States, to eliminate many domestic uses of 2,4,5-T, including homeowner use and use on food crops. Continued uses in forestry, on rangeland, on rights of way, and in rice fields were permitted, resulting in the application of 6 to 7 million pounds of 2,4,5-T in 1974 (Gough, 1986, pp. 137–138).

In 1978 the Environmental Protection Agency, which by then had assumed responsibility for regulating pesticide use in the United States, announced that it was considering cancelling all uses of 2,4,5-T and other herbicides made from trichlorophenol (Gough, 1986, p. 138). This action was motivated by concern about the developmental toxicity reported in the original series of experimental animal studies and was confirmed and extended in subsequent investigations; these included studies in rats that spanned several generations (Murray et al., 1979).

In the spring of 1979, the EPA suspended, on an emergency basis, essentially all remaining uses of 2,4,5-T and related herbicides. The suspension was based on the laboratory data as well as new epidemiology data that purported to correlate herbicide use in Oregon with a seasonal increase of human miscarriages. These studies, involving people residing in the Alsea, Oregon area, were reported widely in the press. At the time, EPA policymakers were persuaded to take the extraordinary action of invoking an emergency suspension based on two current facts: the Alsea study results and knowledge that the start of annual spraying of the forests was imminent. The policymakers were briefed by agency scientists on the adequacy of the epidemiology data and reportedly received similar opinions from a few academic scientists whose cursory opinions were solicited.[1] Detailed analysis of the Alsea results, which occurred during the cancellation process, led most scientists to conclude that there were major flaws in the study that made the authors' conclusions untenable. The toxic effect of concern, at that time, remained focused on developmental abnormalities, as reported in humans and demonstrated in experimental animals. In 1983, after extensive hearings on the merits of permanent cancellation of 2,4,5-T, the last company that held a pesticide registration withdraw from the cancellation proceedings being conducted by the EPA. As a result, a regulatory decision based on the administrative hearings (which would have included a detailed discussion of health risk) was never issued and the pesticide registrations were cancelled through established administrative processes.

Agent Orange

Agent Orange was the name given by the military to a pesticide formulation that was composed of equal parts of the herbicides 2,4,5-T and 2,4-D. At the time Agent Orange was formulated, both herbicides were registered for use in the United States. The military's evaluation of herbicides in Vietnam led to the conclusion that there was significant military benefit associated with their use.

The military use of chemicals as defoliants was a politically sensitive issue because of fears that it would be viewed internationally as a form of chemical warfare. Many Americans were opposed to the war; some American scientists were also vocal in their opposition to herbicide use. Several scientific organizations formally expressed their concern that broad use would cause significant environmental damage (Gough, 1986, pp. 54–55). It has been estimated that up to 10 percent of the land mass of South Vietnam was sprayed with a herbicide by the military (Gough, 1986, p. 49). Herbicide use in Vietnam began in 1962. Because of its effectiveness, Agent Orange became the preferred defoliant in 1965, accounting for 60 percent of the total herbicide used in Vietnam. The peak spraying years were 1967 to 1969, with an estimated total of 10 to 12 million gallons of Agent Orange applied (Gough, 1986, p. 51). The data indicating that a constituent, 2,4,5-T, and the TCDD contaminant could cause birth defects in mice was cited as a major factor in first limiting use of the herbicide in 1969 to areas that were not heavily populated and, finally, ending its use in 1970.

In the mid-1970s a number of veterans sought health care from the VA for psychological and organic illnesses they believed were a consequence of service in Vietnam. It was a nurse in a Chicago VA hospital who first gained public attention by asserting that many of these effects were due to exposure to herbicides used in Vietnam. Organizations representing veterans soon began aggressive lobbying for Vietnam veterans to receive disability compensation for a range of illnesses. A variety of committees were established to review data and advise on these issues. On the recommendation of an advisory committee, the Department of Veterans Affairs (VA) currently compensates veterans that served in Vietnam if they develop chloracne or soft tissue sarcoma. Based on a decision of the Secretary of Veterans Affairs, compensation is also granted to veterans who develop non-Hodgkin's lymphoma. The VA took this action after being briefed by the Centers for

Disease Control that there appeared to be an increase in this disease in servicemen who served in Vietnam; it does not correlate with increased Agent Orange exposure. Currently, there is a proposed regulation that would compensate veterans for peripheral neuropathy that developed within 10 years of the last date of service in Vietnam.

As specified in the Agent Orange Act of 1991 (P.L. 102-4), the Institute of Medicine of the National Academy of Sciences has undertaken a review and evaluation of the available scientific evidence regarding associations between disease and exposure to dioxin and other chemical compounds in herbicides used in Vietnam. The report is to be issued in the latter part of 1993 and will provide scientific and medical information for the Secretary of Veterans Affairs to consider when making determinations about compensation.

Clean Water Act

Section 303 of the Clean Water Act requires states to adopt water quality standards to protect the public health or welfare and enhance the quality of water. States must adopt numeric criteria for all listed pollutants and submit new or revised standards, containing numeric values, to the EPA for review and approval. TCDD is a listed toxic pollutant and is a chemical for which EPA developed and published recommended criteria for human health and aquatic life. The recommended numerical values for water from which fish may be caught for human consumption is 0.014 parts per quadrillion. [If humans also drink the water, the value is 0.013 parts per quadrillion (ppq).] Sixteen states have adopted the EPA recommended value, which is equated with a maximum increased cancer risk of one in one million. In 1990 the state of Maryland proposed a numerical value of 1.2 ppq, almost 100-fold higher than the recommended EPA value. As required by the Clean Water Act, the Maryland proposal was reviewed by EPA and approved on September 12, 1990. In its letter of approval, EPA clearly acknowledged that there are varied choices that can be made in the development of a risk assessment. While some of these choices may differ from those selected by EPA, they are equally defensible from a scientific perspective. In the Maryland proposal, many of the scientific assumptions were the same as those of EPA; however, where they differed, assumptions that had been used by the Food and Drug Administration were used. In addition, Maryland chose to employ a policy that would tolerate an increased cancer risk level of one in one hundred thousand, a decision that is clearly within the prerogative of a state under the Clean Water Act. This policy initiative was facilitated by language in the Clean Water Act that gives the states the right to develop different standards as long as they are scientifically defensible. Five additional states have received approval for numerical values similar to those adopted by Maryland.

THE PUBLIC PERCEPTION OF CHEMICAL RISK

Public policy is not developed in a vacuum nor is it the handmaiden of scientific information. It is made by people who are well aware of the currents and countercurrents of the social issues of the day. Specific decisions are made within a social context. Coincident with the development of experimental data on the toxic effects of TCDD, a series of events occurred that fostered the growing unease that the public was subject to significant chemical exposures over which they had no control and which could result in adverse effects. Consider the following examples:

• The late 1960s and early 1970s was the impact period of Rachel Carson's book, *Silent Spring*, which forcefully and graphically portrayed the consequence of what appeared (and probably was) to be an indiscriminate use of many pesticides.
• The Vietnam War was controversial; spraying chemical defoliants as part of the war was even more so. In 1970 the public learned of the presence of a mystery chemical, TCDD, in sprays that were used to defoliate large areas in Vietnam. Humans were exposed to these sprays; in some instances it may have been our own troops. Laboratory tests of its toxicity were not concluded until after an extensive period of spraying, a fact that seemed to fit the pattern of indiscriminate use described by Rachel Carson.
• In 1974, studies in Missouri revealed that waste oil, applied to dirt roads and horse arenas to suppress dust, contained TCDD and was responsible for sickness and death in horses and possibly accounted for illness in two youngsters (Carter et al., 1975). In 1982 an entire town, Times Beach, Missouri, was purchased by the state and federal government, who stated the action was necessary as a result of TCDD contamination that occurred through the application of waste oil to unpaved streets.
• In the summer of 1976, thousands of people in the town of Seveso, Italy, were contaminated with TCDD when an overpressured reaction vessel in a chemical plant vented its contents to the outside air. Many hundreds of animals became sick, some died (the effects were only partially due to TCDD; many were caused by other chemicals, a fact that was not grasped, and therefore poorly reported by the media). Portions of the town adjacent to the plant were evacuated.
• The 1979 Alsea, Oregon, episode, widely reported on television, brought the issue of aerial spraying "home" in that it was U.S. forests that were sprayed, domestic watersheds that were possibly contaminated with TCDD, and American women who feared that their miscarriages resulted from the spraying. Other uses of 2,4,5-T caused some to wonder if contamination of meat and milk could occur.
• In the early 1980s, companies that had manufactured Agent Orange

agreed to a settlement of a class action suit filed in the federal court. The settlement created a $180-million fund to be shared by Vietnam veterans. From the perspective of the general public, it appeared that chemical companies paid a large sum to settle an issue (Agent Orange) while staunchly maintaining that the chemical did not cause health effects (i.e., the axiom that actions speak louder than words).

DISCUSSION

The saga of TCDD portrays the policy process accommodating scientific and medical information in the context of broader social issues. In some instances, scientific data on adverse effects were persuasive for policymakers and in other instances scientific doubt or uncertainty as to adverse effects resulted in hesitancy to establish or change policy.

Courtney and colleagues reported in 1969 and 1970 that TCDD could cause birth defects in animals. These results had a profound impact on management policy and regulatory decisions. The noteworthy result, from a scientific perspective, was not the nature of the developmental abnormality but the exceedingly low doses of TCDD that could lead to this developmental effect. Thus began the fascination of scientists for studying how TCDD could elicit biological effects that range from birth defects to cancer, immunosuppression, and death. TCDD also causes a potent and persistent stimulation of important enzymes.

The unpublished results of the laboratory tests, showing that the dioxin in 2,4,5-T caused teratogenic effects, were leaked to the press, and soon after a congressional hearing was held. The focus of the hearing was concern that the dioxin in Agent Orange was causing birth defects in Vietnamese civilians. The hearing was well covered by the press, and many members of the public were introduced to information on dioxin and its toxic effects at that time.

The toxicity data indicating potential for developmental abnormalities led to decisions to restrict and subsequently suspend the use of Agent Orange in 1970. The rapid pace of the decision-making process was undoubtedly influenced by the social turmoil associated with the war itself and the belief of some people that herbicides were an insidious form of chemical warfare.

The Agent Orange issue resulted in social pressure to review the broad range of domestic uses of 2,4,5-T in the United States. Again, the scientific data drove the decision to restrict its use but the rapid pace at which the pesticide registrations on many products were revised or cancelled by the Department of Agriculture in 1970 may have been influenced by agricultural interests and pesticide manufacturers who preferred that the issue be resolved before pesticide regulation was transferred to the newly created

EPA. While use of these herbicides by the general population was essentially eliminated, as were many agricultural uses, other agricultural uses that were believed not to involve significant human exposures were retained.[2]

Several years later, scientific data on birth defects again influenced congressional and legal decisions. Specifically, the negative results from the CDC birth defects study persuaded Congress not to compensate Vietnam veterans for children born with birth defects. In the Agent Orange class action suit, brought on behalf of Vietnam veterans against manufacturers, the presiding judge, Judge Jack Weinstein, in summing up the evidence stated, "no laboratory nor epidemiologic evidence exists at this time that is sufficient to link deaths or birth defects to parental exposures to herbicides while serving in Vietnam" (Gough, 1986, pp. 115, 117).

The degree to which scientific information influenced the Veterans Administration and the Congress on the issue of carcinogenicity is not clear. The mandate of the Department of Veterans Affairs is to provide care and compensate for illnesses that are plausibly service connected, a standard that is less rigorous than establishing causality as a result of exposure to a herbicide containing TCDD. Some argued that a causal link needed to be established between the illness and TCDD (Agent Orange). Others postulated that providing medical service, implying that an illness may be related to Agent Orange, was a policy without a scientific base. While scientific certainty is not a prerequisite to the VA policy decision-making process, "granting of medical service" fosters the perception that links have been clearly established.[3]

To the public, the issue of estimating the cancer risks associated with exposure to TCDD remains unsettled. Cancer risk was the dominant health effect cited in the decision to purchase Times Beach, Missouri, an action that is regarded by many as a politically expedient decision of the moment in which TCDD health concerns were subsequently invoked as a rationale. At any rate, this action provides a dramatic contrast to the decision to reoccupy Seveso, Italy, when reduced TCDD levels in much of the city were attained. The levels of TCDD that posed a health risk for Times Beach were estimated using linear multistage models. Thus, these levels were several orders of magnitude lower than if the alternative model, using the NOEL with a safety factor, had been used. Use of the latter model is favored by the Europeans.

In some respects the central issue regarding the estimation of TCDD cancer risk may not be TCDD per se, but EPA's adherence to a method of estimating cancer risks that provides results far more restrictive than required by its current cancer guidelines. The EPA still uses TCDD risk estimates derived from the linearized model. While it did propose a modest revision of the estimate in 1988, it seemingly was thwarted by a scientific peer review that stated there was no new scientific evidence to support a change from the current estimates. In retrospect, the peer review group should have been assigned the additional task of determining if the scientific basis for the *existing* EPA position was any stronger or more compelling than the position that EPA had proposed to adopt. At the same time the reviewers stated that there were no new data to support change, they were critical of the methodology used to develop the existing estimates of risk! There was ample language in the peer review report to permit a policy determination like that which was proposed, that is, increase the exposure level from 0.006 pg/kg/day to 0.1 pg/kg/day.

In response to the peer review, senior policy officials maintained the current standard. A number of months later, the same officials directed the start of a broad reassessment of health and environmental risks associated with exposure to dioxins. That process, which is still going on, gives a significant degree of flexibility in interpretation and judgment to expert scientists drawn from a broad array of interests across the country. This approach contrasts with the traditional practice in which the initial assessment is developed by agency risk assessors, with revisions based on public comment. Observing that a consensus had emerged among expert scientists at a recent conference regarding initial biological events following TCDD exposure, these EPA officials believed that it was an opportune time to revisit current risk assessment practices. The continuing collegial approach encourages a wider dialogue and has the potential to better reflect a broad-based scientific consensus.

During this same period, regulatory action under the Clean Water Act reflected an explicit acknowledgment that alternative assumptions and practices can yield risk estimates different from those espoused by EPA that are of equal scientific plausibility. Constraints in the law may have led the EPA to approve the more permissive state TCDD water quality standards. A broader sentiment supports the view that EPA policymakers seized an opportunity to endorse a mechanism for setting TCDD exposure levels that is less constraining than the existing agency procedure. By doing so, they approved values that are closer to those derived by other federal agencies. This action under the Clean Water Act, as well as prompting the dioxin reassessment process described above, suggests that the EPA's rigid institutional commitment to a narrow modeling approach for assessing cancer risks may have been weakened. If so, EPA may become more receptive to review and use of newer knowledge.

TCDD also appears to be the catalyst for fundamental changes in a number of procedures used to examine scientific data in order to assess health risk and express it quantitatively. The EPA reassessment process, as well as the Institute of Medicine report, which is to be submitted to the VA, is breaking new ground in addressing a long-standing need: a process in which scientific judgments are periodically reviewed and either reaffirmed or revised as necessary. This is not a revolutionary concept; periodic review of national ambient air quality standards was an implicit requirement in the early Clean Air Acts. The fact that this is now occurring may represent a belated evolution in actual practices.

A policy judgment about a chemical risk almost always represents decision making in the face of uncertainty. The principal elements leading to uncertainty are a paucity of data that describe the effect; lack of data that characterize the nature and degree of exposure; and little understanding of the physiological processes involved in the observed effect. In such circumstances, assumptions are made and uncertainty factors incorporated which, in principle, should be replaced with actual data when they become available. In practice, however, the system does this only grudgingly, after concerted effort. It is to be hoped that a process that rewards greater "certainty" of scientific judgment would serve as an incentive for the development of better information.

The dioxin reassessments also venture into issues of chemical additivity and antagonism, and force risk assessments to be broadened to health effects that transcend cancer. They also present an opportunity (some would term it an obligation) to define a more significant role and utility for data on humans. In some respects an attitude seems to have evolved in risk assessment in which laboratory data and model estimates reign supreme. While animal data are the legitimate backbone of preventive toxicology, human data and medical judgment merit a more prominent role. For example, sound human data exist on immune response as a function of body burden of TCDD. There are also experimental data on animals. In instances where both animal and human data yield different estimates of risk, should acceptable levels of exposure be derived from the human data or the animal data? For a number of dioxin risk considerations, this choice will be the crucial policy issue.

From a management and policy perspective, the issues that need to be addressed sometimes are clearer than either science or risk assessment perceives them to be. For example, the animal data indicate that TCDD is a carcinogen. The long-established default assumption has been to treat TCDD as if it were a human carcinogen based on these data. With our existing level of knowledge, this assumption is unlikely to change. A debate continues on the "proper" interpretation of human epidemiology data regarding cancer. From a policy perspective, this is not a major issue; the debate is not going to yield a definitive answer; thus, based on animal data, TCDD will continue to be viewed as a human carcinogen. However, of central importance for both regulatory and VA policy is scientific judgment regarding the dose that may increase the risk of cancer. Thus the extensive epidemiology data on chemical workers assembled by Fingerhut et al. (1991)

may be useful for quantitative estimates. Even if one assumes (and many vehemently oppose this as an assumption) that the relative rates of cancer are related to TCDD exposure, these rates can only be viewed as a relatively weak (low) response in a population where exposure was probably a thousandfold higher than the highest levels occurring today. That is, it is a weak response relative to the high estimate of potency derived from models of animal data. In a more pedestrian way, it certainly could affect the statement commonly uttered that "EPA considers dioxin to be one of the most potent carcinogens studied."

A final area in which risk assessment is changing as a consequence of TCDD is the tacit acknowledgment that data that identify mechanisms by which adverse effects develop can be very helpful in developing more meaningful scientific judgments of risk. Exploring the development and use of biologically based risk assessment models, which is a priority element of the EPA reassessment, is an important change in the philosophic approach to evaluating the risk of cancer. Such changes in the types of scientific data used to assess chemical risk are vital if scientific knowledge is to more accurately inform and enlighten those responsible for social policy.

NOTES

1. The emergency suspension withstood legal challenge in Federal District Court. Under emergency suspension, sale and use ceases immediately; in essence, a product is off the market while the process that leads to a decision to permanently cancel, or otherwise modify conditions of use, is under way. The cancellation process entails the preparation and presentation of documents that contain the data and other information that form the bases for the cancellation. There is an opportunity for review and comment on this document, including administative hearings, before a final decision is issued by the EPA.

2. The statute under which pesticides are registered mandates that regulatory decisions be based on a risk-benefit analysis on the grounds that the mere presence of risk alone is insufficient to prohibit use. The risks of use must be weighed against the benefits to be realized from use.

3. The process of reconciling policies with current scientific knowledge about TCDD continues, as exemplified by the requirements of the Agent Orange Act of 1991, described earlier. Legislation calling for a National Academy of Sciences report on current scientific understanding underscores the continuing unease and uncertainty regarding the scientific facts as well as the desire that the data and scientific judgments may guide a VA policy that fairly responds to the health concerns and afflictions of Vietnam veterans.

REFERENCES

Air Force Health Study. 1991. Serum dioxin analysis of 1987 examination results. Epidemiology Research Division, Brooks Air Force Base, Texas.

Barnes, D., A. Alford-Stevens, L. Birnbaum, F. W. Kutz, W. Wood, and D. Patton. 1991. Toxicity equivalency factors for PCBs? Quality Assurance: Good Practice Regulation and Law 1(1):70–81.

Carter, C. D., R. D. Kimbrough, J. A. Liddle, R. E. Cline, M. M. Zack, and W. F. Barthel.

1975. Tetrachlorodibenzodioxin: An accidental poisoning episode in horse arenas. Science 188:738–740.

Centers for Disease Control Veterans Health Study. 1988. Serum 2,3,7,8-tetrachlorodizenzo-p-dioxin levels in U.S. Army Vietnam-era veterans. JAMA 260:1249–1254.

Collins, J. J., M. E. Strauss, G. J. Levinskas, and P. R. Conner. 1993. The mortality experience of workers exposed to 2,3,7,8-tetrachlorodibenzo-p-dioxin in a trichlorophenol process accident. Epidemiology 4:7–13.

Courtney, K. D., D. W. Gaylor, M. D. Hogan, H. L. Falk, R. R. Bates, and I. Mitchel. 1970. Teratogenic evaluation of 2,4,5-T. Science 168:864–866.

Courtney, K. D., and J. A. Moore. 1971. Teratology studies with 2,4,5-trichlorophenoxyacetic acid and 2,3,7,8-tetrachlorodibenzo-p-dioxin. Toxicology and Applied Pharmacology 20:396–403.

Erickson, J. D., J. Mulinare, P. W. McClain, T. G. Fitch, L. M. James, A. B. McClearn, and M. J. Adams, Jr. 1984. Vietnam veterans' risks for fathering babies with birth defects. JAMA 252:903–937.

Ericksson, M., L. Hardell, and H. O. Adami. 1990. Exposure to dioxins as a risk factor for soft tissue sarcoma: A population based case control study. Journal of the National Cancer Institute 82:486–490.

Ericksson, M., L. Hardell, N. O. Berg, T. Moller, and O. Axelson. 1981. Soft-tissue sarcomas and exposure to chemical substances. A case-referent study. British Journal of Industrial Medicine 38:27–33.

Evans, R. G., K. B. Webb, A. P. Knutsen, S. T. Roodman, D. W. Roberts, J. R. Bagby, W. A. Garrett Jr., and J. S. Andrews, Jr. 1988. A medical follow-up of the health effects of long term exposure to 2,3,7,8-tetrachlorodibenzo-p-dioxin. Archives of Environmental Health 43:273–278.

Fingerhut, M. A., W. E. Halperin, D. A. Marlow, L. A. Piacitelli, P. A. Honchar, M. H. Sweeney, A. L. Greife, P. A. Dill, K. Steenland, and A. J. Suruda. 1991. Cancer mortality in workers exposed to 2,3,7,8-tetrachlorodibenzo-p-dioxin. New England Journal of Medicine 324:212–218.

Gough, M. 1986. Dioxin, Agent Orange: The Facts. New York: Plenum Press.

Gough, M. 1988. Science policy choices and the estimation of cancer risk associated with exposure to TCDD. Risk Analysis 8:337.

Gough, M. 1991. Editorial: Agent Orange: Exposure and Policy. American Journal of Public Health 81:289–290.

Hardell, J. L. 1981. On the relation of soft tissue sarcoma, malignant lymphoma, and colon cancer in phenoxy acids, chlorophenols and other agents. Scandinavian Journal of Work Environment and Health 7:119–130.

Hardell, L., M. Ericksson, P. Lenner, and E. Lundgren. 1981. Malignant lymphoma and exposure to chemicals, especially organic solvents, chlorophenols and phenoxy acids: A case control study. Scandinavian Journal of Work Environment and Health 43:169–176.

Hardell, L. and A. Sandstrom. 1979. Case-control study: Soft-tissue sarcomas and exposure to phenoxyacetic acids or chlorophenols. British Journal of Cancer 39:711–717.

International Agency for Research on Cancer. 1972. IARC Monographs on the Evaluation of Carcinogenic Risk of Chemicals to Man, Volume 1. International Agency for Research on Cancer, World Health Organization, Lyon.

Kahn, P. C., M. Gochfeld, M. Nygren, M. Hansson, C. Rappe, H. Velez, T. Ghent-Guenther, and W. P. Wilson. 1988. Dioxins and dibenzofurans in blood and adipose tissue of Agent Orange-exposed veterans and matched controls. JAMA 259:1661–1667.

Kang, H. K., K. K. Watanabe, J. Breen, J. Remmers, M. G. Conomos, J. Stanley, and M. Flicker. 1991. Dioxins and dibenzofurans in adipose tissue of U.S. Vietnam veterans and controls. American Journal of Public Health 81:344–349.

Kimbrough, R. D. 1990. How toxic is 2,3,7,8-tetrachlorodibenzodioxin to humans? Journal of Toxicology and Environmental Health 30:261–271.

Kociba, R. J., D. G. Keyes, J. E. Beyer, R. M. Carreon, C. E. Wade, D. A. Dittenber, R. P. Kalnins, L. E. Frauson, C. N. Park, S. D. Barnard, R. A. Hummel, and C. G. Humiston. 1978. Results of a two-year chronic toxicity and oncogenicity study of 2,3,7,8-tetrachlorodibenzo-p-dioxin in rats. Toxicology and Applied Pharmacology 46:279–303.

Lamb, J. C. 4th, T. A. Marks, B. C. Gladen, J. W. Allen, and J. A. Moore. 1981a. Male fertility, sister chromatid exchange, and germ cell toxicity following exposure to mixtures of chlorinated phenoxy acids containing 2,3,7,8-tetrachlorodibenzo-p-dioxin. Journal of Toxicology and Environmental Health 8:825–834.

Lamb, J. C. 4th, J. A. Moore, T. A. Marks, and J. K. Haseman. 1981b. Developmental and viability of offspring of male mice treated with chlorinated phenoxy acids and 2,3,7,8-tetrachlorodibenzo-p-dioxin. Journal of Toxicology and Environmental Health 8:835–844.

Lathrop, G. D., W. H. Wolfe, R. A. Albanese, and P. M. Moynahan. 1984 (unpublished). An epidemiologic investigation of health effects in Air Force personnel following exposure to herbicides. Baseline morbidity study results. U.S. Air Force School of Aerospace Medicine, Aerospace Medical Division, Brooks Air Force Base, Texas.

Lathrop, G. D., W. H. Wolfe, S. G. Machado, J. E. Michalik, and T. G. Karrison. 1987 (unpublished). An epidemiologic investigation of health effects in Air Force personnel following exposure to herbicides. First follow-up examination results. U.S. Air Force School of Aerospace Medicine, Aerospace Medical Division, Brooks Air Force Base, Texas.

Manz, A., J. Berger, J. W. Dwyer, D. Flesch-Janys, S. Nagel, and H. Waltsgott. 1991. Cancer mortality among workers in chemical plant contaminated with dioxin. The Lancet 338:959–964.

Mastroiacovo, P., A. Spagnolo. E. Marni, L. Meazza, R. Bertollini, G. Segni, and C. Borgna-Pignatti. 1988. Birth defects in the Seveso area after TCDD contamination. JAMA 259:1668–1672.

Mocarelli, P., A. Marocchi, P. Brambilla, P. Gerthous, D. S. Young, and N. Mantel. 1986. Clinical laboratory manifestations of exposure to dioxin in children: A six year study of the effects of an enviromental disaster near Seveso, Italy. JAMA 256:2687–2695.

Murray, F. J., F. A. Smith, K. D. Nitschke, C. G. Humiston, R. J. Kociba, and B. A. Schwetz. 1979. Three-generation reproduction study of rats given 2,3,7,8-tetrachlorodibenzo-p-dioxin in the diet. Toxicology and Applied Pharmacology 50:241–252.

Nygren, M., C. Rappe, G. Lindström, M. Hansson, P. A. Bergqvist, S. Marklund, L. Domellöf, L. Hardell, and M. Olsson. 1986. Identification of 2,3,7,8-substituted polychlorinated dioxins and dibenzofurans in environmental and human samples. Chlorinated Dioxins and Dibenzofurans in Perspective. Chelsea, Mich.: Lewis Publishers.

Rappe, C., P. Bergqvist, M. Hansson, L. Kjeller, G. Lindström, S. Marklund, and M. Nygren. 1984. Chemistry and analysis of polychlorinated dioxins and dibenzofurans in biological samples. Banbury Report.

Schecter, A., J. D. Constable, J. V. Bangert, H. Tong, S. Arghestani, S. Monson, and M. Gross. 1989. Elevated body burdens of 2,3,7,8-tetrachlorodibenzodioxin in adipose tissue of United States Vietnam Veterans. Chemosphere 18:431–433.

Schultz, K. H. 1957. Klinische und experimentelle Untersuchungen zur Ätiologie der Chlorakne. Archiv fuer Klinische und Experimentelle Dermatologie 206:589–596.

Selected Cancers Cooperative Study Group. 1990. The association of selected cancers with service in the U.S. Military in Vietnam. I. Non-Hodgkin's Lymphoma. Archives of Internal Medicine 150:2473–2483.

Smith, A. H., D. G. J. Patterson Jr., M. L. Warner, R. MacKenzie, and L. L. Needham. 1992.

Serum 2,3,7,8-tetrachlorodibenzo-p-dioxin levels of New Zealand pesticide applicators and their implication for cancer hypotheses. Journal of the National Cancer Institute 84:104–108.

Sparschu, G. L., F. L. Dunn, R. W. Lisowe, and V. K. Rowe. 1971. Study of the effects of high levels of 2,4,5-trichlorophenoxyacetic acid on foetal development in the rat. Food and Cosmetics Toxicology 9:527–530.

Tognoni, G., and A. Bonaccorsi. 1982. Epidemiological problems with TCDD (a critical view). Drug Metabolism Reviews 13:447–469.

U.S. Environmental Protection Agency. 1988. A cancer risk-specific dose estimate for 2,3,7,8-TCDD. External review draft. EPA/600/6-88/007.

Webb, K. B., R. G. Evans, and A. P. Knutsen. 1989. Medical evaluation of subjects with known body levels of 2,3,7,8-tetrachlorodibenzo-p-dioxin. Journal of Toxicology and Environmental Health 28:183–193.

Wolfe, W. H., J. E. Michalek, and J. C. Miner. 1991. Air Force Health Study. An epidemiologic investigation of health effects in Air Force personnel following exposure to herbicides. Serum Dioxin Analysis of 1987 Examination Results. Epidemiology Research Division, Armstrong Laboratory, Human Systems Division (AFSC), Brooks Air Force Base, Texas.

Woods, J. S., S. L. Polissar, R. K. Severson, L. S. Heuser, and B. G. Kulander. 1987. Soft tissue sarcoma and non-Hodgkin's lymphoma in relation to phenoxy herbicide and chlorinated phenol exposure in Western Washington. Journal of the National Cancer Institute 78:899–910.

Zeiger, E. 1989. Genetic toxicity. Pp. 227–238 in Halogenated Biphenyls, Terphenyls, Naphthalenes, Dibenzodioxins and Related Products, 2nd ed. , R. D. Kimbrough and A. A. Jensen, eds. New York: Elsevier.

Zober, A., P. Messerer, and P. Huber. 1990. Thirty-four-year mortality follow-up of BASF employees exposed to 2,3,7,8-TCDD after the 1953 accident. International Archives of Occupational and Environmental Health 62:139–157.

Keeping Pace with Science and Engineering. 1993.
Pp. 243–250. Washington, DC: National Academy Press.

Science, Engineering, and Regulation

Richard D. Morgenstern

The principal problem posed in this volume is that the policy and regulatory process does not incorporate new scientific information very well or in a timely manner. No matter how good or poor the state of understanding was about the costs, risks, or benefits associated with a particular environmental issue, sooner or later that understanding is likely to change, presumably for the better. When it changes, the new understanding may call an earlier decision into question. In light of this inevitability, we should be concerned that the regulatory system is sufficiently flexible to be able to adapt.

In thinking about this issue, we should recognize that all regulations are not created equal. Depending on the situation, different levels of scientific evidence are required to promulgate a regulation. Driven by statutory mandates and deadlines, tight budgets, and other directives, a type of "triage" system is applied to different environmental problems.

The case studies in this volume—all of which deal with changes in the information base—suggest that there are three distinct but related circumstances in which the U.S. Environmental Protection Agency (EPA) may be called upon to revisit decisions:

1. The original decision was based on the best science at that time, but subsequently the science changed.

---

*The author is indebted to Frederick Allen, Devra Lee Davis, Albert McGartland, and Myron F. Uman for invaluable comments.

2. The agency was pressed to make decisions in such a short time frame or under such a tight budget constraint that it was not possible to evaluate the science thoroughly. Economic and other pressures have now raised concerns about the scientific validity of the original decision.

3. The original decision was based solely on the availability of pollution control technology, and more recent evidence has brought to light some environmental or public health problems that were not previously foreseen.

The rationale for change is the clearest when the original decision was based on the best science then available and now there are new data suggesting that the decision should be revisited. Particularly in cases where the science is making significant new advances, we clearly need to incorporate the new information into our regulations and policies.

In this volume, the case study on dioxin comes closest to this description. As described in the case study, EPA is in the process of reconsidering its risk assessments of dioxin, with an eye to what EPA should do next. The agency has gone through a public process but in the interim has committed to enforcing the regulations until any changes are made.

Chlorofluorocarbons (CFCs) are another good example. The best scientific data available were brought to bear when the Montreal Protocol was drafted in 1987. The drafters knew, however, that the science was rapidly evolving and wisely wrote in a provision to the protocol that allows for revisions based on new scientific information when it becomes available. In 1990, based on such evidence, the 50 percent reduction in CFCs that had been agreed to just three years earlier was transformed into a phaseout by the year 2000, and several additional ozone-depleting chemicals were added to the agreement. Two years later, in 1992, the phaseout date was moved up to 1994 and several additional chemicals were added to the agreement.

The second category involves situations where, often because of the pressure for quick action, the original decision was not always based on the most thorough understanding of the science. An example here is the body of drinking water regulations. Congress charged EPA with promulgating a large number of regulations in a short time but did not correspondingly expand the resources to accomplish the task. Now, state and local governments face increased financial burdens to comply with these regulations. The key question here is how to remedy the situation? In 1992 the National Governors' Association convened several high-level meetings and ultimately issued a series of recommendations that called for slowing the implementation of certain drinking water regulations. After considering a moratorium on implementing already promulgated regulations, Congress enacted the Chafee-Lautenberg Amendment to the EPA Appropriations Act, which mandated a major study on the implementation of the Safe Drinking Water Act, including analyses of the costs and benefits, state implementation, and small systems issues. Whether this will ultimately lead to significant programmatic or statutory changes remains to be seen.

Some Superfund remedies fall into this category too. In a number of instances states have raised concerns about the wisdom of the proposed site remedies, 10 percent of the costs of which are typically assumed by the states themselves. This issue will inevitably be considered in the reauthorization of the Comprehensive Environmental Response Compensation and Liability Act.

The third category, in which the original decision was based solely on the availability of pollution control technologies, is illustrated by the case study on the municipal waste combustor New Source Performance Standard. In that instance, EPA had the flexibility to regulate under either Section 111 or 112 of the Clean Air Act. Using the latter would have involved a risk-based determination. Based on its discretion, the agency chose, instead, to issue the regulations under the authority of Section 111, which applies a technology-based approach to controlling emissions. In a similar vein, the standards for hazardous waste treatment generally rely heavily on engineering solutions, often ignoring a balancing of risks, benefits, and costs.

THE LAWS AND THE REGULATORY PROCESS

Some provisions of the laws EPA administers require the agency to revisit decisions periodically. The most notable example is the National Ambient Air Quality Standards—the engine of the Clean Air Act—which, by statute (Section 109), are to be reconsidered every five years. In fact, these reconsiderations have consistently taken longer than five years. Fourteen years elapsed between the last two formal decisions on the ozone standard and, in the end, the standard was simply reaffirmed. The primary SO_2 standard was revised in 1979 and it is still being reconsidered.

Another example is the reregistration of pesticides. EPA is required under the 1988 amendments to the Federal Insecticide, Fungicide, and Rodenticide Act to reregister the approximately 700 active ingredients within 10 years. Many of these active ingredients were registered when risk assessment methodology was extremely primitive, and the data requirements for registration were minimal. Because of this, there are many registered active ingredients that have not undergone rigorous human or ecological risk assessments. To date, the EPA has reregistered only a handful of active ingredients; however, the pace should pick up.

When originally enacted, many technology-based standards, such as the New Source Performance Standards (Section 111) of the Clean Air Act, were intended to be technology forcing. The paint industry, for example, has reduced its rate of emissions of volatile organic compounds (VOCs) by more than 20 percent by substituting water-based solvents for petroleum-based solvents, largely as a result of the technology-based emission limits for that industry.

Technology-based standards, however, can also inhibit innovation. Standards are often based on a particular technology, which at the time may be considered to be at the cutting edge. Permit writers, knowing that the standard setters relied on that technology to determine the limit, rarely allow compliance by the use of other technologies except after rigorous (often costly) demonstrations by the applicant. Thus the technology may become entrenched in the industry and there is little incentive to innovate or to achieve emissions beyond those required by the standard.

Most of the nation's environmental laws allow EPA to reconsider its decisions at its discretion—in other words, it is up to EPA. Critics note that the agency does not have a strong record for meeting deadlines for the reevaluation of regulatory decisions. Often this is due to the budget constraints under which EPA operates and pressures on the agency to address as yet unregulated environmental problems.

Another factor to consider is that the EPA's regulatory process is typically analytical, inclusive, and very public—especially when compared with the procedures used in other countries. In contrast to most parliamentary systems, the U.S. government is organized around a system of checks and balances on the theory that adversarial tensions are required to ensure fair and balanced outcomes. The Administrative Procedures Act mandates a complex set of steps, including analysis, development of regulatory options, formal proposal, public comment, reassessment, and final rule. The key participants in the process include EPA, the regulated community, environmental organizations, the media, and the research and engineering communities.

The case studies in this volume show how our open, democratic process must accommodate not only scientific concerns, but also political, legal, and institutional issues. Many legal scholars would undoubtedly argue that the process itself is as important as the regulatory outcome.

THE ROLE OF SCIENTISTS AND ENGINEERS

Science, engineering, and their practitioners should be, and in many cases are, an integral part of the development, drafting, and implementation of regulations. In practice, scientists and engineers assist in ensuring that regulations are based on the best technical understanding available. That is, they help to translate technical concepts into implementable, enforceable, realistic regulations. Clearly, many would argue for strengthening the role of scientists and engineers in the regulatory process.

More technically defensible decisions inevitably lead to better allocation of the resources available to be devoted to environmental protection. At the same time, reconsidering decisions involves costs that can be reduced if we get it right the first time. Changing regulations creates uncertainty and a less stable business environment. Since EPA has severely limited resources, revisiting past decisions often means other work is delayed.

As Robert M. White notes in his introduction to this volume (p. 6):

> . . . the environmental regulatory system should . . . keep pace with the changes in our understanding of the technical aspects of the issues, and it should remain stable on a time scale sufficient for regulated parties to comply with some measure of economic efficiency. It is evident that these two normative characteristics can be, and frequently are, in conflict.

In this regard, the most important thing that scientists and engineers involved in the regulatory process can do is to understand policymakers' information needs. Scientists and engineers should consider the following:

1. Quantify and measure outcomes that have meaning in a policy context. What does a reduction in forced expiratory volume mean to decision makers? How does that translate into lifestyle or health changes that people understand? Similarly, hazards research that is too general and fails to develop dose-response relationships is often difficult to use.

2. Don't be afraid to make best judgments. Decisions often have to be made under tight deadlines. It is usually better that judgments about science and engineering be made by scientists and engineers rather than by laymen. EPA often uses the National Research Council or its own Science Advisory Board to get best judgments about tough issues. The Clean Air Scientific Advisory Committee (CASAC) has played a key role in helping the agency to set ambient air standards. This type of mechanism should be expanded to encourage scientists and engineers to make best judgments on unresolved technical issues.

3. Work on the right technical issues. This is perhaps the hardest problem of all to solve. Good research and development usually involves a long lead time and the EPA process often cannot provide sufficient notice to get it going. However, just as EPA can, to some extent, shape its agenda to reflect the work of researchers in the field, so could the technical experts take EPA's needs into account as they set their research priorities. EPA's regulatory needs should not be the only determinant of the research and development agenda, but its needs should play a significant part in shaping that agenda. To help establish research agendas for both intramural and extramural research, EPA has begun an interactive consultation process involving program managers and scientists inside the agency as well as experts from outside.

WHAT OTHER INSTITUTIONS CAN DO

I have covered some things that EPA and the science community might do to make use of new scientific information in the regulatory process. Let me turn to the issue of how other key institutions can become helpful catalysts for change as well.

1. *The mainstream media.* Both print and broadcast media can provide more complete coverage of technical issues relevant to environmental regulation. A number of newspapers and journals are beginning to acknowledge that environmental policy cannot be based just on what Senator Moynihan has called "middle-class enthusiasms." The media should focus more on facts and not just perceptions. Scientists and engineers can help reporters by taking the time to speak clearly and frankly to the media. In many technical quarters media coverage is shunned. Yet scientists lose their right to complain about scientific illiteracy if they do not themselves contribute to a more informed public.

2. *Congress.* The *Washington Post* reported that the House Space, Science, and Technology Committee, formerly obscure, was the second most popular pick for freshman members in 1993, largely because of its potential for pork barrel politics (*Washington Post*, January 26, 1993). In fact, pork barrel spending does the double harm of funding often unneeded programs and forcing aside programs founded on good technical assessments of scientific and policy priorities. In fiscal year 1993 the Congress added more than 100 specific items to the EPA budget while approving essentially flat spending levels relative to 1992. Of course, in such circumstances, something has to give. Many programs of national interest, based on scientific risk assessment, were cut to make room for these congressionally mandated programs, many of which had not been subject to technical review and did not carry high technical priority.

3. *Environmental groups.* As environmental groups become better versed in technical issues, their contributions to the debate in science-based policy questions are becoming more useful. As many as 15,000 environmental groups currently exist in this country, most at the community level. It would be unrealistic to expect many of these groups to be dispassionately concerned with technical analysis. Some of the national groups, such as the Nature Conservancy, are becoming more conversant in scientific issues and are using science as part of their own priority setting. The Environmental Defense Fund is gaining expertise in both economic and science issues. But a staff study done at the EPA in 1992 indicated that science most often comes into the loop in these organizations after the membership, board, or staff decides what problems to target. Clearly, further progress is called for.

Building a commonly accepted science base will enhance trust and help ensure that we are not working at cross-purposes.

4. *The regulated community.* Finally, the regulated community can fund more high-quality research and development. The most effective arguments are based on good technical and economic analyses. Too often industry channels its resources to lobbying efforts at the expense of collecting real data or even contributing relevant data already in its possession. EPA has established procedures for accepting proprietary data and has now a solid track record with such data. One can see the dawning of a more progressive and involved view in many parts of the business community, but clearly the time has come for industry to make a larger contribution to developing relevant scientific and engineering information and making it readily available to EPA.

DISCUSSION

The way a problem is posed often dictates its solution. Thus, the problem considered in this volume of case studies is how to improve the use of scientific and technical information in the regulatory process. The solutions presented in the case studies generally offer a variety of approaches regarding more and better scientific information. As described above, enhanced mechanisms to improve the use of technical information in the regulatory process are clearly needed.

It is possible that this focus on better data and analysis has some downsides as well. Two types of risks need to be examined—the risk of falsely regulating something that needs no such control, a false positive, and the risk of falsely exonerating something that poses a hazard.

While many observers argue that the costs of false positives are quite high, at least one analyst, Carl Cranor (1993) of the University of California, suggests that the costs of false negatives—particularly in the environmental field—are significantly greater than the costs of false positives. Research scientists are characterized as aspiring to find more and better data, to withhold judgment until sufficient empirical information is obtained, to be cautious about inferences about causation, to be ever skeptical regarding conclusions, and to guard against basing decisions on false positives, that is, evidence suggesting something is a problem, when it, in fact, is not.

This type of "better science," as Professor Cranor notes, "when . . . used inappropriately or insensitively . . . can produce substantial bad consequences as well as good." To reduce these undesirable consequences, he argues for expanded administrative discretion regarding scientific and technical information:

A better approach would be to address uncertainties by policy choices . . .

determine the minimum amount of information needed to guide decisions and then use it, develop rebuttable interim standards for protecting public health, find an appropriate balance between false positives and false negatives for the context, expedite the evaluation of known carcinogens [and other hazards], and increase the rate at which substances are identified as toxins (Cranor, 1993, p. 103).

While one can dispute the frequency of false negatives—it may be most relevant where the data on risk and other aspects are particularly lacking—Congress clearly thought that the problem of air toxics emissions represented a false negative; Congress dropped the risk approach in favor of a technology standard as an initial basis for regulating a specified list of pollutants. Even acknowledging such examples of false negatives, however, it is probably not the principal problem in the field of environmental protection. Yet, it is based on a cogent assessment of the complex and changing nature of environmental sciences. Certainly we should be mindful of the issues raised by such an approach as we consider our future paths.

REFERENCE

Cranor, Carl F. 1993. Regulating Toxic Substances: A Philosophy of Science and the Law. New York: Oxford University Press.

Keeping Pace with Science and Engineering. 1993.
Pp. 251–262. Washington, DC: National Academy Press.

Environmental Regulation and Technical Change: Overview and Observations

J. Clarence Davies

The case studies in this volume are a rich trove of information related to environmental regulation and technical change. They provide insight into a wide variety of situations where the policy community and the scientific and engineering community meet or fail to meet.

The cases present a spectrum of types of decisions, the most important distinction being between determining what problems should be considered and how to deal with a particular problem. The cases present a variety of enviromental problems, ranging from formaldehyde to acid rain and from trihalomethanes to vegetation in Chesapeake Bay. Across the cases, there are many different actors, each with different values, views, and roles. The cases deal with decisions at the state and local level as well as with decisions made by the federal government. The state of scientific understanding relevant to each decision varies widely, from a relatively well understood problem like acid rain to a situation like disinfecting by-products in drinking water where half the by-products are still unidentified. In all cases, the relevant technical data usually suffer from some degree of uncertainty and therefore are subject to interpretation.

Despite the multitude of differing factors between the case studies and within each one, it is possible, if difficult, to draw some tentative generalizations and make some general observations triggered by the case studies.

INCORPORATING NEW TECHNICAL INFORMATION

The problem that was uppermost in the minds of the Steering Committee when it commissioned the case studies was the extent to which environmental regulations take account of new scientific discoveries. This is a basic problem but, as will be discussed later in this paper, not the only type of problem illustrated by the cases.

The cases present a broad range of the times elasped between the discovery of new relevant technical information and its incorporation into regulating decisions. Only five years elapsed between the discovery of trihalomethanes in drinking water and regulations designed to reduce their levels, but in most cases it took more time to incorporate new information. The Chesapeake study estimates the typical lag between discovery and action at "ten years or more" (p. 29). The ozone study estimates that the lag is "five to fifteen years" (p. 79).

The cases cite many different factors that influence how much time it takes to readjust. Three seem to be of particular importance: the extent to which the new information threatens the status quo; the degree of uncertainty of the scientific data; and the involvement of the public or legislators in the decision.

All other things being equal, the greater the threat posed by new information to the status quo the longer it takes to incorporate that new information into decisions. The status quo can be defined both for the public sector and for the private sector. The public sector status quo consists of all the previous decisions made by the government. The process of making regulatory decisions typically consumes a good deal of time, money, and political capital. Government regulators are therefore loath to upset decisions that have already been made. The private sector status quo is established by the technological and other investments that the regulated parties made to comply with previous policy decisions. Naturally, the firms that have made such investments are reluctant to see the regulatory basis for the investment changed in any way.

Both the public and private sectors regard new technical information in the context of the capital—political as well as financial—that has been sunk into past decisions, and both may be reluctant to make new investments to accommodate new information. As between the public and the private sector, it seems likely that the degree of threat to the public sector will generally be more important in accounting for the time it takes to incorporate new information, although the case studies do not provide good evidence on this.

The operational corollary to the status quo hypothesis is that the least delay will occur in regulating new problems because these represent less of a threat to the status quo. An example is formaldehyde, which had not been regulated (except for acute effects) before the Chemical Industry Institute of Toxicology (CIIT) rat study. Formaldehyde as a carcinogen was placed high on the regulatory agenda even before the CIIT study was complete, and it was regulated almost immediately by the Consumer Product Safety Commission (CPSC). The CPSC regulation was subsequently overturned by the Court of Appeals, the court stating that CPSC had erred in relaying solely on the CIIT study.

The longest delays in incorporating new information will occur in relation to methodological information that threatens a large number of previous decisions. Research on dioxin and later scientific findings about formaldehyde threaten to require a change in the way the federal government models exposure-response relations for carcinogens. In turn, such a change in modeling threatens all of the standards that have previously been established for carcinogens. As of this writing, the federal agencies have not accepted any change in the way that they model carcinogenicity, although a large-scale review of dioxin toxicity is under way.

The degree of threat to the status quo may be the most important factor in explaining the interval between the discovery of new information and incorporating it in regulatory decisions, but the degree of certainty of the new scientific data and how the data are communicated are also important variables. In general (and not unexpectedly), the more uncertain the decision maker is about the validity of the new information, the longer will be the delay in using the data.

Much of the regulatory action (or inaction) described in the case studies takes place against a background of scientific disagreement. Whether it is the interpretation of the formaldehyde epidemiological studies, the dioxin animal data, or the health risk from chlorination, a high degree of uncertainty characterizes the state of knowledge. In a case such as dioxin, the primary disputes seem to be between one set of scientists and another rather than between scientists and regulators. In this context the Chesapeake case documents an interesting shift of initiative from managers to scientists as the complexity of protecting the Bay from excess nutrients was increasingly recognized. Although scientific uncertainty may increase the discretion exercised by regulators, in the cases involving compliance with ambient standards (Chesapeake Bay, tropospheric ozone) the complexities of the sources, transport, and effects of various pollutants may lead to increasing reliance on models and on the scientists who construct and run the models.

A third significant variable in explaining the time required to adjust regulations to new understanding is the degree of public or legislative involvement. However, the relationship between this variable and the length of time to incorporate information is not straightforward. A high degree of involvement seems to produce more rapid regulatory action, but regulators may overreact and the action may be too rapid or the involvement too intense to allow use of the best available scientific information. This seems to have been the case, for example, in the proposal for recycling requirements as a condition for approving new municipal waste combustors. Other examples not covered in the cases readily come to mind.

Delay does not seem to be influenced by whether the new information would make a regulation more stringent or less. In the dioxin and formaldehyde cases, information that would make regulation less stringent has not been incorporated. But there have also been long delays in using new information about tropospheric ozone precursors that would result in more stringent controls, and the same is true of controls for limiting nitrogen inputs in Chesapeake Bay. The formaldehyde case presents us with a situation where the regulated industry, confronted with a choice between less stringency and the status quo, chooses the latter.

ACTING WITHOUT ADEQUATE SCIENCE

Delay in incorporating new technical information can result in regulatory action (or inaction) based on bad science, but regulators may take actions poorly supported by science because information is simply not available or is deliberately ignored.

The case studies implicitly use two quite different models of the regulatory process. One model characterizes decision makers as reactive—they delay doing anything until new information pushes them into action. This model highlights the problem of delay in incorporating new information. However, the second model characterizes decision makers as more impulsive. In this model decision makers need to move ahead, need to make decisions, and they will use whatever science—good or bad—is available at the time. The authors of the ozone case study describe this problem succinctly: "Information sufficient to resolve key technical uncertainties seldom becomes available within the time frame in which policymakers feel compelled to act. . . . '[A]n adequate knowledge base'—that required to support truly informed decision-making—is the pot at the end of the rainbow: while we may try, we never seem to get there" (p. 85).

Two general factors account for this problem. The first is that decision makers are subject to an array of forces that compel them to take action—statutory deadlines, court mandates, pressure from legislators or constituents. These forces can frequently lead to regulatory timetables that are unconnected to the process of developing needed information. The municipal waste combuster case provides a good example of such pressures at work.

The other factor involves the limitations on the science base for making environmental decisions and the uncertainties inherent in our technical understanding of environmental problems. Almost every scientific discipline has some relevance to environmental problems, but environmental problems are relatively new concerns for almost every discipline. Even when adequate resources have been made available, it simply takes a long time for new technical information to be developed and verified. Thus, as each of the cases demonstrates, decision makers frequently must operate with a major gap between what they need to know and what the technical community can tell them.

An interesting and important situation arises when the technical understanding is available but is incompatible with the existing regulatory framework. For example, it is well established scientifically that nitrogen oxides simultaneously contribute to the problems of tropospheric ozone, acid rain, and visibility. However, the regulatory process does not take account of this simultaneity and treats nitrogen oxides separately in the context of each problem because each problem is addressed through a separate regulatory program, which in turn is authorized by separate legislative provisions. It could be argued that the regulatory approach of dividing the world of chemicals into two sharply distinct categories—carcinogens and noncarcinogens—is also a result of the need for government agencies to try to simplify a complex world.

POLICY AGENDA NOT BASED ON SCIENCE

If actions unsupported by science and engineering are taken because the regulatory agenda is unrelated to the scientific agenda, it becomes necessary to investigate how the agendas are established. The case studies shed a good deal of light on this question, and they also highlight two sets of additional problems: the extent to which the political agenda is not based on science and the extent to which the scientific agenda is distorted by politics or is not responsive to the needs of policymakers.

Two major recent EPA reports, *Unfinished Business* and *Reducing Risk*, concluded that the enviromental problems that receive the most attention are not the environmental problems that pose the greatest risks to public health or the environment.[1] The political agenda—the issues that receive priority attention from government officials—has little relationship to an agenda based on good technical information about risks.

The case studies prepared for this symposium do not directly address the priority-setting problem, but they do contain numerous insights into how the political agenda is established. They suggest that it is established in a two-step process. First, some new technical information appears; then the new information receives a push from some political force—an interest group, a congressional committee, the President. The lesson is that new technical information by itself does not significantly influence the political agenda—it must be assisted by some type of political propellant.

The case studies contain examples in which scientific information by itself has had some influence. For example, the acid rain case states (p. 171), "The public release of . . . three reports with prestigious scientific imprimaturs was a major reason the Reagan administration felt compelled to initiate limited planning for a national strategy to reduce acid deposition and shifted away, at least symbolically, from exclusive reliance on its requirement for further research."[2] However, the fact that the Reagan shift was only symbolic may support the two-step hypothesis.

The Chesapeake Bay case suggests that physical events can influence the political agenda. Here, Hurricane Agnes led to the Chesapeake Bay study, which in turn led to political action. Whether a dramatic physical event can substitute for political sponsorship or whether the dynamics of agenda-setting are different at the regional level than at the federal level are unanswered questions. The major role played by the media in interpreting both physical events and technical findings also must be considered.

The state of scientific understanding also influences the political agenda by serving as a reality check. In a broad sense, of course, the policy-maker's picture of physical reality is largely determined by scientific and engineering information, although the information may be highly filtered or actually erroneous by the time it reaches the policymaker. A more concrete example of this function of technical information is given in the ozone study, where government agencies were forced to face a variety of policy questions after empirical studies showed that automobile inspection and maintenance systems did not reduce automobile emissions to the degree that had been predicted.

Technical information, or its absence, also can keep issues off the political agenda. For example, even though long-range transport of ozone was required to be considered under the provisions of the Clean Air Act, it was not considered or regulated because of the lack of "specific data suitable for air quality modeling and analysis of long-range transport" (p. 68).

THE INFLUENCE OF POLITICS
ON THE SCIENTIFIC AGENDA

Just as technical information influences the political agenda, so the reverse is true. The scientific agenda is composed of the subjects and types of research undertaken by the technical community. Politics affects the state of technical understanding through the authorization and funding of research and testing programs. Political influence on the scientific agenda at times can be excessive or insufficient. If it is insufficient, the scientific agenda will be nonresponsive to policy needs and the technical information needed by decision makers will be unavailable. If it is excessive, necessary research may not be done, new problems may go unidentified, and the integrity of the scientific process will suffer.

The responsiveness (or lack thereof) of the scientific community to policy needs was brought sharply to attention by the very large investment made in acid rain research under the National Acid Precipitation Assessment Program (NAPAP). The acid rain case study concludes that NAPAP had some but not much impact on the 1990 Clean Air Act. The author reminds us that, "the political process, not science per se, dictates how much information is enough as well as the conditions under which information guides a given policy decision" (p. 183). Nevertheless, to the extent that NAPAP failed to provide an adequate technical basis for the 1990 amendments to the Clean Air Act there is clearly a problem because this is the purpose for which the NAPAP was established.

On a smaller and less visible scale, the NAPAP experience may be a common occurrence. The author of the ozone case comments, "Excessive uncertainty . . . may be a consequence of governmental research programs not being sufficiently long term and consistently focused to provide the information needed to reduce uncertainties to acceptable levels" (p. 83). The Chesapeake Bay case contains a variant of this problem dealing with large-scale quantitative models: "As the cost of the model increases (in terms of time and money) and the corporate memory is lost, the model begins to take on a life of its own and the predictions become reality. Thus, there is a tendency for the management community to reach the conclusion that additional scientific information is no longer needed." (p. 32).

Two types of political impact on the science and engineering agenda are very important but are not dealt with extensively in the case studies. First is the neglect of monitoring and studies to support the evaluation of government programs. Government officials tend to place a low priority on evaluative studies and this is reflected in the science agenda. For example, the ozone study notes that VOC and NO_x levels are not measured and that, "This paucity of information severely limits the ability to evaluate the effectiveness of emissions control programs" (p. 77).

The second type of impact, one on which the Steering Committee had hoped to collect information, is that of regulations on the development and adoption of new technologies. The case studies do not shed much light on this question, and what limited light they shed is different in different cases. The authors of the ozone case study state, "regulations generally prompted the development of technologies required for compliance" (p. 86). However, in the municipal waste case (p. 131), the Clean Air Act standards based on "best demonstrated technology" prevented EPA from forcing the development of new technology. Similarly, the Chesapeake Bay case (p. 30) notes that "Clearly, reliance on a particular technology (secondary treatment) as the basis for regulating nutrient inputs has inhibited the development of alternative (less costly, more effective?) approaches and technologies."

One can conclude from these observations in the case studies that performance standards (those that just establish the level of compliance to be achieved) encourage new technology while standards that specify a particular type of technology are an impediment to the diffusion, and perhaps to the development, of new technologies. Such a conclusion may well be valid, but the whole area of the relationship between regulation and technology deserves more detailed and intense exploration.

IMPROVING THE PROCESS

The authors of the case studies provide a number of suggestions for improving the way the regulatory process relates to scientific and engineering information. Before reviewing these suggestions, it is worth keeping in mind Robert White's basic point that the goal of any improvements must be achieving a balance between often competing forces.

In his introduction to this volume, White states (p. 5–6), ". . .as desirable as it is to have regulations that are based on the best, most current technical understanding, it is also desirable to have a stable regulatory regime within which the affected parties can intelligently plan to come into compliance and implement their plans. . . . We have then two characteristics that we all would agree the environmental regulatory system should exhibit: it should respond dynamically to changes in our understanding of the technical aspects of the issues, and it should remain stable on a time scale sufficient for regulated parties to comply with some measure of economic efficiency. It is evident that these two normative characteristics can be, and frequently are, in conflict."

A related point that needs to be made is that the regulatory process involves many considerations besides "science" and "politics." A Manichean view of the process that sees it as a contest between the bad politicians and the good scientists is no more helpful than a view that puts all the emphasis on changing to adapt to new information or a view that only emphasizes stability. There are many other factors that must be taken into account when evaluating or analyzing the regulatory process—economics, problems of implementation, legal requirements, public opinion, to name just a few. The process is characterized by numerous actors, and recommendations to improve it must consider multiple values.

More research. There can be little dispute that the general state of knowledge regarding environmental problems is often woefully inadequate. The case studies are, in one sense, simply a series of examples supporting this general statement. Furthermore, the missing knowledge is sometimes fairly elementary—for example, the effect of recycling requirements on emissions from municipal incinerators. An obvious conclusion from the cases is that more research is needed. Because enviromental regulations tend to involve information at the cutting edge of technical understanding, more research is almost always desirable.

This conclusion should be accompanied by the caveat, also illustrated by the cases, that more research may not always help in a particular situation. The massive epidemiological study of formaldehyde conducted by Blair, involving 600,000 person-years of data, did not help much in settling the controversy over formaldehyde. Similarly, many of the studies of dioxin did little to settle that controversy.

Several participants in the symposium raised the question of what type of institution should do the necessary research. The basic point made was that the short-term demands of the regulatory process provide an inhospitable context for conducting long-term high-quality research, and that therefore it may be useful to separate organizationally the regulators from the long-term programs. Although not raised in the discussion, ensuring that technical information is incorporated into the regulatory process is a competing value. The closer the researchers are to the regulators organizationally, the more likely it is that their research results will be incorporated in regulatory decisions. Balancing these values and exploring organizational innovations to reconcile them is high on the list of issues that warrant further examination.

Communication. Several of the cases comment on the importance of communication between scientists and engineers on the one hand and decision makers on the other. For example, the authors of the Chesapeake Bay study state (p. 31) that the lack of communication (in addition to other differences such as differing time perspectives) had the result that "(1) the management community tends to question the relevance of environmental research conducted by an independent science community, and (2) the science community tends to question the integrity of the management process."

The communication problem takes several distinct forms. Communication may not take place at all; information may be communicated but become distorted; or it may be communicated and understood but ignored by the decision maker. Ways of dealing with communication problems may help in all these situations or may help in dealing with only one of them.

In-house capacity. It seems likely that a basic variable in facilitating communication is the extent to which the regulatory agency possesses an in-house scientific capability. If no one in the agency understands the technical issues, communication with the technical community obviously will be difficult.

Although the federal and state agencies covered by the cases varied widely in their scientific capabilities, the cases themselves do not emphasize this difference. It is interesting that in the formaldehyde case the agency that responded most rapidly to the first evidence of formaldehyde carcinogenicity was the Consumer Product Safety Commission, the agency with the least in-house technical capability. There may be times when in-house capability is an impediment to action. However, it is also relevant that the CPSC decision was overturned in the courts for lack of scientific support.

Science advisers are used by many agencies to facilitate communication of technical information. Richard Morgenstern, the acting deputy adminis-trator of EPA, observes elsewhere in this volume (p. 247) that the type of science advisory mechanism that has served EPA in the past "should be expanded to encourage scientists and engineers to make best judgments on unresolved technical issues."

Peer review is another mechanism to both facilitate communication and provide some quality control over the technical information used to make policy decisions. The EPA Science Advisory Board for a number of years has provided some peer review of the science used in regulatory decisions, and the agency recently has moved to stricter requirements for peer review of scientific data.

The cases seem to indicate that most of the regulatory agencies make little use of various formal techniques for dealing with uncertainty or other aspects of decisions. In effect, they seem to lack in-house capability in the decision sciences despite the fact that they are repeatedly faced with the situations and problems for which the methods of decision science were developed.

The intergovernmental dimension. In the U.S. federal system, one fac-tor affecting the relationships between the technical community and the regulators may be the level of government that makes the regulatory deci-sion. The cases contain a good deal of information about interactions among the different levels of government but, typically, the cases deal with differ-ent types of interactions in different circumstances and suggest different conclusions.

In the municipal waste combuster case, the authors conclude: "Our findings suggest that the type of technical and scientific information used may be different at each level of government . . . local regulatory agencies have the most to gain by seeking and requiring the most stringent air pollu-tion control and the highest standards for operation of facilities, regardless of actual risks associated with these facilities" (pp. 128–129). The authors of the Chesapeake Bay case note the same disregard of scientific data but draw an opposite conclusion with respect to stringency: "In contrast to the perspective of federal officials and reports by local scientists and citizens' groups, state officials in Maryland insisted that the Bay was doing just fine . . . . Thus, the governing body responsible for the implementation of nutrient control plans, the State, was least receptive to scientific evidence indicating the early stages of Bay-wide eutrophication" (p. 15). In both cases, roles are reversed at different stages—sometimes the federal government is most receptive to new information and urges the most stringent standards, sometimes the state is most receptive and stringent.

The authors of the ozone case study observe that, "where California has assumed a leadership role, it seems to have encountered fewer barriers to action, acceptance, and implementation than has the federal government" (p. 86). It seems doubtful that state officials in Maryland or Virginia responsible for the Chesapeake Bay would agree. About the only certain conclusion that can be drawn is that the intergovernmental dimension is an important variable in understanding the regulatory process.

Courts. Surprisingly, the cases deal very little with the role of the courts, although many environmental regulatory decisions are finally decided by litigation, as are the schedules for administrative action. The courts have their own difficulties dealing with technical information, and in recent years experiments have been performed using scientific advisers to judges, panels of expert technical witnesses that can be used by a judge, and other methods to facilitate the objective use of technical knowledge in the courtroom setting.

Sequencing. Robert White notes that, "We need to build into the structure of the regulatory system means for reconsidering earlier decisions if and when our understanding changes sufficiently to call earlier decisions into question" (p. 5). The same theme is sounded in the ozone case (p. 85): "Where circumstances require action, such as smog conditions in the South Coast Air Basin, the waiting time for research results exceeds the time practically available for taking the action(s). Resolution of this dilemma is exceedingly difficult. One option is to design actions that can be carried out in sequence, instituting more stringent controls with time, as needs warrant. The results of research can then influence the 'action sequence' as they become available."

Policies can be structured to allow the introduction of new technical information. Although the requirement that National Ambient Air Quality Standards be reviewed every five years has not been adhered to, nevertheless it has served as a prod for the regulators to consider new technical findings. The provisions of the Vienna Convention on Stratospheric Ozone have worked effectively to adjust regulatory requirements rapidly to changing scientific understanding.

Regulatory negotiation. Two of the case studies suggest that formal negotiation among interested parties to develop a regulation can facilitate incorporation of technical information. The authors of the municipal waste combustor case state, "If negotiations were to become a standard component of a regulatory process, stakeholders would have a more controlled and focused opportunity to provide EPA with information in an atmosphere of cooperation" (p. 135). However, the cases do not actually show whether this is true in practice. Regulatory negotiation was not used in the combustor case. It was used to deal with a rule on the by-products of water disinfection, but the case study was completed before the accomplishments of the negotiation could be evaluated.

Several other groups are examining proposals to improve the use of technical information in environmental regulation and policy. Organizations such as the National Academies of Sciences and Engineering are a continuing presence in this area while groups such as the Carnegie Commission on Science, Technology, and Government and the World Wildlife Fund's National Commission on the Environment have recently focused on the connection between technical information and policymaking. The attention that the issue is receiving should result in concrete steps toward improving the science-policy relationship.

CONCLUSION

Robert White writes in his introduction: "The primary question of policy is this: Does the current environmental regulatory system strike an appropriate balance between dynamic change and stability?" (p. 6). The case studies generally indicate that the answer may be 'no,' although as I have tried to show, the question is in reality very complex and does not lend itself to simple answers.

As important as any overall judgment is an understanding of how the regulatory process actually deals with technical information. Such analysis leads to other questions. For example, the authors of the formaldehyde case state: "Perhaps the ultimate issue . . . is who decides when new science should be incorporated into the regulatory process"—agency risk assessors, risk managers, external scientific panels, Congress, the courts? (p. 217) The case studies provide a good starting point for considering this and related questions. They move us away from rhetoric and toward an empirical base for addressing some of the most important underlying questions of environmental policy.

NOTES

1. See U.S. Environmental Protection Agency, Office of Policy Analysis, *Unfinished Business* (Washington, D.C., February 1987); and U.S. Environmental Protection Agency, Science Advisory Board, *Reducing Risk* (Washington, D.C., September 1990).

2. Editor's note: The three reports referred to here are William A. Nierenberg et al., *Report of the Acid Rain Peer Review Panel* (Washington, D.C., Office of Science and Technology Policy, 1984); National Research Council, *Atmosphere-Biosphere Interactions: Toward a Better Understanding of the Ecological Consequences of Fossil Fuel Combustion* (Washington, D.C., National Academy Press, 1981); and U.S. Congress, Office of Technology Assessment, *Acid Rain and Transported Air Pollutants: Implications for Public Policy* (Washington, D.C., Government Printing Office, 1984).

Biographical Data

Steering Committee

CHARLES R. O'MELIA is professor of environmental engineering and chairman of the Department of Geography and Environmental Engineering at the Johns Hopkins University. He received his B.C.E. (1955) from Manhattan College and his M.S.E. (1956) and Ph.D. (1963) in sanitary engineering from the University of Michigan in Ann Arbor. Dr. O'Melia was elected to the National Academy of Engineering in 1989. His research interests are in aquatic chemistry, water and wastewater treatment, and modeling of natural surface and subsurface waters.

J. CLARENCE (TERRY) DAVIES is director of the Center for Risk Management at Resources for the Future. His previous positions have included assistant professor of public policy at Princeton University, executive vice president of the Conservation Foundation, and assistant administrator for policy at the Environmental Protection Agency. Most recently, he served as executive director of the National Commission on the Environment. Dr. Davies is a political scientist who, over the past 30 years, has written several books and numerous articles about environmental policy.

ROBERT C. FORNEY is a retired executive vice president, member of the board of directors, and member of the executive committee of E.I. du Pont de Nemours & Company. Dr. Forney held positions of increasing responsibility in Du Pont, including product manager, director of the Products Marketing Division, general director of the Marketing Division, and vice president and general manager of the Textile Fibers Department. He is a former member of the Board of Governors of the Purdue Foundation and serves as a director on several boards. Dr. Forney received his B.S. and Ph.D. in chemical engineering from Purdue University.

ROGER O. McCLELLAN, D.V.M., serves as president of the Chemical Industry Institute of Toxicology in Research Triangle Park, North Carolina. A member of the Institute of Medicine, he has previously chaired the National Research Council (NRC) Committee on Toxicology and concurrently serves as a member of the NRC Committee on Risk Assessment Methodologies for Hazardous Air Pollutants. He is a former chairman of the EPA Clean Air Scientific Advisory Committee, is concurrently a member of the EPA Science Advisory Board Executive Committee, and has been a member of numerous other advisory groups in government, academe and private industry. He has a long-standing interest in integrating data from human, laboratory animal, and *in vitro* studies to assess human risks from exposure to radiation and chemicals.

M. GRANGER MORGAN is professor and head of the Department of Engineering and Public Policy at Carnegie Mellon University. His research interests include public policies in which technical issues play a central role and techniques for dealing with uncertainty in quantitative policy analysis. He was educated at Harvard University and received his Ph.D. in applied physics from the University of California at San Diego in 1969. Dr. Morgan has served on a number of committees of the National Research Council and EPA's Science Advisory Board.

PAUL R. PORTNEY is vice president and senior fellow at Resources for the Future, an independent, nonpartisan research and educational organization concerned with natural resources and the environment, where he previously was director of its Center for Risk Management. He is also a visiting lecturer at Princeton University's Woodrow Wilson School of Public and International Affairs. In 1979-1980, he served as chief economist at the Council on Environmental Quality in the Executive Office of the President. Dr. Portney received his Ph.D. in economics from Northwestern University and is the author or coauthor of a number of journal articles and books, including *Public Policies for Environmental Protection*.

JOHN H. SEINFELD is the Louis E. Nohl Professor and chairman of the Division of Engineering and Applied Science at the California Institute of Technology. A member of the faculty of Caltech since 1967, he was appointed executive officer for chemical engineering in 1973 and became Louis E. Nohl Professor in 1980. He has been chairman of the Division of Engineering and Applied Science since 1990. Dr. Seinfeld is a member of the National Academy of Engineering and a Fellow of the American Academy of Arts and Sciences. His research interests are in the atmospheric chemistry and physics of air pollution. He received a B.S. from the University of Rochester in 1964 and a Ph.D. from Princeton University in 1967, both degrees in chemical engineering.

MYRON F. UMAN is assistant executive officer for special projects of the National Research Council (NRC). Concurrently, Dr. Uman is a member of the adjunct faculty of George Mason University and a visiting scholar at the Johns Hopkins University. His Ph.D. is from Princeton University (1968) in electrical engineering and plasms physics. At the NRC since 1973, he has managed or conducted more than 20 formal studies of the application of scientific and technical information to the development and implementation of public policy across a wide range of enviromental issues.

Authors

WALTER R. BOYNTON received his B.S. in biology from Springfield College in 1959, M.S. in marine sciences from the University of North Carolina at Chapel Hill in 1974, and Ph.D. in environmental engineering from the University of Florida in 1975. Dr. Boynton's expertise is in estuarine nutrient dynamics and seagrass ecology. He has served on a number of advisory boards and committees, including Science Advisory Board, Maryland Department of Natural Resources, Chesapeake Bay Research and Monitoring Division, Scientific and Technical Advisory Committee to EPA's Chesapeake Bay Program, and the Calvert County Environmental Commission. Dr. Boynton is currently a professor at the Chesapeake Biological Laboratory of the Center for Environmental and Estuarine Studies.

JAMES D. FINE is an environmental analyst with LSA Associates, Inc. Mr. Fine performs air quality, environmental acoustics, and economic studies related to municipal development and natural resource management issues. Mr. Fine joined LSA after working as an information systems analyst at Anderson Consulting. Mr. Fine holds a B.S. degree in economics from the University of Pennsylvania's Wharton School of Business (1989).

MICHAEL GOUGH is manager of the Biological Applications Program at the Office of Technology Assessment. He received his Ph.D. degree at Brown University in 1966. After a 10-year academic career, including two Fulbright Lectureships, he first joined OTA in 1978. He directed projects in environmental and occupational health in the OTA Health Program and OTA's oversight programs related to Agent Orange and atomic bomb test veterans.

Between 1985 and 1990, he was a consultant and director of the Center for Risk Management, Resources for the Future. He is the author of *Dioxin, Agent Orange* (Plenum, 1986) and coeditor with T.S. Glickman of *Readings in Risk* (Johns Hopkins, 1990).

JOHN D. GRAHAM is professor of policy and decision sciences at the Harvard School of Public Health, where he teaches the methods of risk analysis and benefit-cost analysis. Dr. Graham is the founding director of the Harvard Center for Risk Analysis, which promotes analytical thinking about societal responses to health, safety, and environmental hazards. He also heads the Harvard Injury Control Center, which promotes science-based interventions to control trauma from both intentional and unintentional causes. He is the author of four books and dozens of scientific articles and serves on the international editorial boards of *Risk Analysis* and *Accident Analysis and Prevention*.

THOMAS W. HORTON received his bachelor's degree in liberal arts with a concentration in economics from the Johns Hopkins University in 1968. He is a writer and a naturalist. From 1972 to 1987, Mr. Horton was a reporter for the Baltimore Sun, specializing in environmental topics. He has received regional and national awards for environmental coverage of Chesapeake Bay. In 1987 he published *Bay Country*, a collection of essays that won the 1988 John Burroughs Medal of the Museum of Natural History in New York. In 1988 Mr. Horton wrote *Turning the Tide*, the first comprehensive assessment of the state of the Chesapeake Bay. His most recent book on the Chesapeake, *Waters Way*, was published in 1992. Currently Mr. Horton writes a weekly column on "Environment" for the *Baltimore Sun* and is working on another book.

JASON E. JOHNSTON is a graduate of the Massachusetts Institute of Technology, where he studied in the Technology and Public Policy Program and the Department of Chemical Engineering. Mr. Johnston is currently a staff specialist with Karch and Associates, Inc., a scientific consulting firm specializing in toxicology, epidemiology, and risk assessment. Mr. Johnston has experience in evaluating exposures to, and the associated health risks of, toxic substances in industrial settings and at hazardous waste sites. He has also performed analyses of environmental data and information in the environmental policy, regulatory compliance and litigation arenas.

RENATE D. KIMBROUGH is senior medical associate at the Institute for Evaluating Health Risks. She is a doctor of medicine with training in pathology, diplomate of the American Board of Toxicology, and fellow of the American Academy of Clinical Toxicology. Dr. Kimbrough conducted research in toxicology and environmental health at the Centers for Disease Control for 27 years before being appointed adviser to the administrator of the U.S. Environmental Protection Agency for Medical Toxicology and Risk Evaluation in 1989. She joined IEHR in 1991.

THOMAS C. MALONE received his B.A. in zoology from Colorado College in 1965, M.S. in oceanography from the University of Hawaii in 1967, and Ph.D. in biology from Stanford University in 1971. Dr. Malone's expertise is in plankton dynamics and the processes of eutrophication in coastal ecosystems. He has served on a number of advisory boards and committees, including the executive board of the Scientific and Technical Advisory Committee to EPA's Chesapeake Research Consortium, and vice chair of the University National Oceanographic Laboratory System Advisory Council. Dr. Malone is the director of the Horn Point Environmental Laboratory of the Center for Environmental and Estuarine Studies and of the Multiscale Experimental Ecosystem Research Center, an EPA Center for Exploratory Environmental Research.

JOHN A. MOORE is president and chief executive officer of the Institute for Evaluating Health Risks (IEHR). IEHR is established as a nonprofit institution to serve government, industry and the public on issues that address the health risk of chemicals. Before joining IEHR, Dr. Moore was assistant administrator of the Office of Pesticides and Toxic Substances at the U.S. Environmental Protection Agency. He came to EPA from the National Institute for Environmental Health Sciences, where he was both director for toxicology research and testing and deputy director of the National Toxicology Program. Dr. Moore received a Doctor of Veterinary Medicine degree from Michigan State University in 1963; he is also certified by the American Board of Toxicology.

RICHARD D. MORGENSTERN holds a Ph.D. in economics from the University of Michigan (1970). He was a tenured associate professor of economics at Queens College of the City University of New York before becoming the deputy assistant director for energy, the environment, and natural resources at the Congressional Budget Office in 1977. Subsequently, he served as legislative assistant to Senator J. Bennett Johnston (1979–1980) and then the director of the Energy Program of the Urban Institute. Since 1982, he has been director of the Office of Policy Analysis of the U.S. Environmental Protection Agency. While at EPA, he has served as acting assistant administrator for policy, planning, and evaluation (1991–1993) and as deputy administrator during the transition period at the beginning of the Clinton administration.

SUELLEN WERNER PIRAGES received her Ph.D. at Stanford University. Her area of expertise is biological sciences, with emphasis on environmental toxicology and genetics. Dr. Pirages was a senior staff officer in the Environmental Studies Board of the National Research Council during 1977 to 1890. She then served as a senior analyst in the Industry, Technology, and Employment Program of the U.S. congressional Office of Technology Assessment. From 1984 to 1990, Dr. Pirages was managing director for environmental policy and director of hazardous waste programs for the National Solid Waste Management Association. Currently, Dr. Pirages is executive director at Karch & Associates, Inc., a scientific consulting firm specializing in toxicological and epidemiological evaluations, regulatory and legislative policy analyses, and risk assessment applications.

SUSAN W. PUTNAM is a research associate in environmental policy at the Harvard Center for Risk Analysis. She received her doctoral degree from the Department of Health Policy and Management at the Harvard School of Public Health in 1991. Since joining the Center for Risk Analysis, Dr. Putnam has continued her doctoral research exploring the role of scientific advisory groups in public health policymaking and has examined the risks and benefits of chlorinated drinking water.

JAMES L. REGENS is Freeport-McMoRan Professor of Environmental Policy at Tulane University in New Orleans, Louisiana. While with the U.S. Environmental Protection Agency from 1980 to 1983, Dr. Regens chaired the Group on Energy and Environment of the Organization for Economic Cooperation and Development from 1981 to 1983 and was EPA Joint Chair of the National Acid Precipitation Program from 1981 to 1982. He is vice-chairman of the U.S. Army Corps of Engineers Environmental Advisory Board.

PHILIP M. ROTH is a principal of Envair, an unincorporated association dedicated to carrying out contract research and offering consulting services in the environmental and earth sciences. Dr. Roth has been an independent consultant in environmental science and policy since 1983. He was vice president of Systems Applications, Inc., served as a member of the board of directors, and was technical director and director of environmental studies during the period from 1969 to 1983. Dr. Roth holds B.S. and M.S. degrees in chemical engineering from the Massachusetts Institute of Technology and a Ph.D., also in chemical engineering, from Princeton University.

PHILIP C. SINGER is a professor in the Department of Environmental Sciences and Engineering at the University of North Carolina at Chapel Hill, where he is also director of the Water Resources Engineering Program. He has conducted research on the chemical aspects of water and wastewater treatment for the past 28 years, for the past 17 years focusing on the formation and control of disinfection by-products in drinking water. Dr. Singer is a past chairman of the Research Division of the American Water Works Association. He is currently on the editorial board of the journal *Ozone Science and Engineering* and was formerly an associate editor of *Environmental Science and Technology*. He is also a member of the National Research Council's Water Science and Technology Board. His S.M. and Ph.D. degrees in environmental sciences and engineering are from Harvard University.

COURT STEVENSON received his B.S. in biology from Brooklyn College of the City University of New York in 1966 and his Ph.D. in 1972 from the University of North Carolina at Chapel Hill. He began research in wetlands at the University of Maryland at College Park in 1972 where he was among the first to recognize the massive decline of submersed aquatic vegetation in Chesapeake Bay and link it to nutrient loadings from the surrounding watershed. His current research interests include diffuse source inputs at the land-sea interface to seagrass ecology in tropical lagoons, as well as in the impacts of sea-level on coastal systems throughout the world.

ROBERT M. WHITE is president of the National Academy of Engineering and vice chairman of the National Research Council, the principal operating agency of the National Academy of Engineering and the National Academy of Sciences. Dr. White has had a distinguished career in environmental science and engineering. He established one of the first corporations devoted to environmental science and services. He served in the government under five Presidents, from 1963 to 1977, first as chief of the U.S. Weather Bureau, and finally as the first administrator of the National Oceanic and Atmospheric Administration. Prior to his election as president of the NAE, Dr. White was president of the University Corporation for Atmospheric Research. He holds a B.A. degree in geology from Harvard University and M.S. and Sc.D. degrees in meteorology from the Massachusetts Institute of Technology.

STEPHEN D. ZIMAN is a staff scientist in the Air Issues and Technology Team, Environmental Group, at Chevron Research and Technology Company. His principal responsibilities are in technical and regulatory areas of air quality planning at the federal and state levels. Dr. Ziman has worked as a research scientist with Chevron Chemical Company and for Chevron USA Environmental Affairs and Chevron Production prior to his present position. He was an American Chemical Society Congressional Science Fellow during 1979-1980, and worked as a staff member for the U.S. House of Representatives Subcommittee on Science, Research, and Technology. Dr. Ziman received his B.S. degree in chemistry from the University of Michigan in 1967 and his Ph.D. in organic chemistry from the University of Wisconsin in 1971.

Abbreviations

BACT	Best available control technology
BAT	Best available technology
BDT	Best demonstrated technology
BMP	Best management practice
BOD	Biological oxygen demand
CAA	Clean Air Act
CARB	California Air Resources Board
CMSA	Consolidated metropolitan statistical area
CPSC	Consumer Product Safety Commission
CWA	Clean Water Act
DBPs	Disinfection by-products
DPX	Delivered dose
EKMA	Empirical kinetic modeling approach, in reference to ambient air quality
EPA	Environmental Protection Agency
ESP	Electrostatic precipitator
FGD	Flue gas desulfurization
FIP	Federal implementation plan
FTP	Federal test procedure
GCP	Good combustion practices
HCHO	Formaldehyde (chemical symbol)
I/M	Inspection and maintenance, in reference to motor vehicle emissions
ISR	Indirect source review

LAER	Lowest achievable emission rate
LEV	Low emission vehicle
LOAEL	Lowest observable adverse effects levels
MCL	Maximum contamination level
MWC	Municipal waste combustion
NAAQS	National Ambient Air Quality Standards
NAPAP	National Acid Precipitation Assessment Program
NOAEL	No observed adverse effect level
NOM	Natural organic material
NSF	National Science Foundation
NSPS	New source performance standard
NSR	New source review
OSHA	Occupational Health and Safety Administration
OSTP	Office of Science and Technology Policy
OTA	Office of Technology Assessment, U.S. Congress
PSD	Prevention of significant deterioration, in reference to ambient air quality
RACT	Reasonably available control technology
RAF	Reactivity adjustment factor
RCRA	Resource Conservation and Recovery Act
RFP	Reasonable further progress
RR	Resource recovery
SAV	Submerged aquatic vegetation
SIP	State implementation plan
SNCR	Selective noncatylic reduction
STP	Sewage treatment plant
TCM	Transportation control measure, in reference to ambient air quality
THMs	Trihalomethanes
TLEV	Transitional low emission vehicles
TOX	Total organic halide, in reference to disinfection by-products in drinking water
TSCA	Toxic Substances Control Act
UAM	Urban Airshed Model
UFFI	Urea-formaldehyde foam insulation
VOC	Volatile organic compound
WTE	Waste to energy, in reference to combustion of solid municipal wastes

Index

C

D